MY CHILDREN AND MITO

MY CHILDREN AND MITO

ISBN 978-1-4303-2374-7

Table of Contents

RANDI, BROOKE AND DREW THERE IS NO GREATER LOVE THAN
THE LOVE I HAVE FOR EACH OF YOU

-Mom

INTRODUCTION

Mito is an abbreviation often used when referring to mitochondrial diseases. Relatively recently mitochondrial diseases have been recognized and diagnosable by the medical community, as a group of very complex chronic conditions. This is why there are very few doctors familiar with these diseases, and why so many individuals are still misdiagnosed or go undiagnosed. Only if you're lucky to find, or be referred to, one of these doctors can an individual be diagnosed and receive appropriate medical care.

Through the years, numerous doctors encouraged me to write a book detailing my family's struggles with this medical mystery. Our family initially faced trying to find answers and help when they didn't exist. Five years ago, after more than nineteen years of literally searching every day, we were diagnosed with mitochondrial disease. All I ever wanted were the answers to my family's growing and complicated medical problems. I had accepted there may not be any further help available. It's been a long, hard, uphill battle to validate that I'm not crazy, and this is

REAL! I hope to spare others suffering with mitochondrial diseases the hellish ordeal we endured by empowering them to overcome obstacles they face, and at the same time offer them the hope they so desperately need.

Mitochondria are small organelles located in nearly all of our bodies' cells and are referred to as "power plants". Mitochondria produce, through a number of very complicated chemical processes, the majority of the energy needed by the body to function and maintain life. This is why mitochondrial diseases can affect all of the body's life sustaining systems.

With today's available technology, numerous forms of mitochondrial diseases have been identified. It is felt there are many more. Current available testing methods can't always detect the complexities and variations of mitochondrial diseases making concrete proof sometimes elusive.

Mitochondrial diseases are genetic and can affect generations of a family. When an entire family or a number of members of the same family are affected, though they may share some similarities, they will all be affected differently. Since it is common for multiple family members to be afflicted with this chronic often debilitating condition, it takes a huge toll on a family. Mitochondrial DNA, inherited only from the mother, is the most common way mitochondrial diseases occur, but not the only means. In some individuals it's evident from birth, as it was in my children. Other times it appears later in life. The multiple symptoms and medical problems arising from mitochondrial diseases are extremely variable from one person to another. Symptoms can fluctuate from minute to minute for no apparent reason and can range from mild to life threatening. It can make such activities as eating, digesting, talking, sitting, walking, breathing, brain functioning and even keeping your eyelids open taxing or impossible at times. Because of the lack of predictability, the majority of the time there is no way to give a prognosis. Individuals don't know how, or if, it may impact their life expectancy.

Currently a cure does not exist. Just as each individual is affected differently by mitochondrial diseases, each individual's response to the multiple treatment options is different. Quite often, many combinations of medications are needed. While none completely resolve all symptoms, many help to some degree. Most treatments targeting the faulty mitochondria consist of a combination of special diets, vitamins, minerals, and supplements. Most are not covered by health insurance companies, Medicaid, or Medicare, because they're not considered prescription drugs. However, these treatments can be every bit as effective as

prescription medications and are needed to maintain normal body functions. In addition, other standard therapies and treatments may also be employed to address multiple medical problems which can stem from mitochondrial diseases.

What little valuable energy a person does possess is depleted in never ending efforts to fight for medical care, educate, and wage battles with insurance companies, agencies, and the system.

I am passionately driven by the determination to make a difference and help turn the tide for all who suffer from mitochondrial diseases and their families. It is hoped my book will help bring to the world a widespread, improved awareness of mitochondrial diseases.

Now, please let me share with you our family's life with MITO.

CHAPTER 1

THE FIRST DAY OF SCHOOL

It was the first day of school, September 7, 1994. Along with it came all of the associated things. We were prepared. Our shopping was complete, new sneakers, clothes, crayons, notebooks, pencils, pens and paper. Butterflies fluttered in our stomachs with anticipation of the new people and places we would encounter today. It was up early, get ready, and out the door. The van was packed, we were on our way...but it wasn't to school. It was supposed to be the first day of fifth grade for all three of my children, - Randi, age twelve; Brooke, age eleven; and Drew, almost ten. Instead, we were off on another 'medical trip'.

This is the phrase we use to describe the trips we must regularly make in an ongoing effort to find answers and help for my children's multiple, chronic medical problems. It seems like we have spent half of our lives in doctors' offices and hospitals since their births.

This time it was off to a new, distant hospital, Johns Hopkins in Baltimore, Maryland. It would take us eight hours one way, if everything went as planned and we didn't have automotive troubles or get lost. I was very nervous about traveling to a big new city, all alone with three sick kids. For one reason or another, no one could go with us this time to help. This particular medical trip was different from all the rest in the past. Compelled by the love for my children, I knew I had to do this no matter what. I couldn't let my apprehensions interfere. A dreadful impending feeling time was running out obsessed me.

Boston Children's Hospital, where the children had gone for the past seven years, had given up on them. Approximately a dozen specialists told us they'd done every test known to mankind, and there was nothing more anyone could do. Doctor's recommended I find a good psychologist and learn to live with their unexplainable illness. How could I learn to live with all three of them deteriorating before my eyes? The amount of daily pain and suffering was horrific! Watching your child suffer, and feeling there's nothing you can do to help is extremely anguishing for a parent. Determined, I swore to myself, no matter what the cost financially, physically, or mentally, I would not stop until I had exhausted every avenue twice over. As I drove, my mind repeatedly reviewed everything I needed to discuss with the doctors. This had become common practice prior to every doctor appointment. Randi, Brooke, and Drew's lives and well-being depended upon it. I feared if I left out any information, it could make a difference in finding the answers and help we so desperately needed. These were my precious innocent children. It was up to me, and me alone, to advocate for them. Generally, children of this age are complacent and stoic. Mine were no exception. The parent sees the daily misery, but children don't complain to doctors as an adult might.

Randi, like many children, just internalized everything and appeared content dealing with a horrible monster that she assumed everyone dealt with. Because she had been born this way, she knew no difference. Randi was a delightful, gentle, quiet, reserved child who seldom opened up to others. Her facial expressions alone were usually her way of communicating with non family members. Strangers easily recognized Randi as a special child due to her unique facial appearance and gait. To outsiders Randi may have given the impression she was lower functioning mentally than she actually was. Randi is a thinker and wise beyond her years in many respects. This was later revealed through her writing.

Brooke, who by this time had outgrown most of what I referred to as her "wild child years", was bubbly and always had a big smile on her face. This was just as deceiving as Randi's quietness. Brooke was personable, caring, very outgoing, and social. She loved to watch others and interact with them every chance she got. This was Brooke's strength.

My son Drew acted like a little man. He was well mannered, respectful, very confident, and capable in everything he did. He analyzed everything and interacted with adults well. He still had a mischievous side, trust me, but all it took was "the look" from mom to put the fear in him and straighten him around. To this day, he will tell you he was traumatized by "the look."

On this white knuckled trip I was soothed by the newly released <u>LION KING</u> sound track the kids listened to repeatedly as I was driving. Somehow, we made it to the hospital in one piece. With only minutes to spare, we hurried to the registration desk.

While standing in line, we surveyed the unfamiliar surroundings. They were not bright and cheery like the children's hospitals we were accustomed to. The cold, black metal and ghostly, opaque white dividers felt strange, even eerie.

We first met with the chief of the G.I. (gastroenterology) department, along with a group of others from his division. It was a lengthy appointment. We had to discuss three children's long, complex, medical histories and summaries. These appointments were always very taxing. Many days have been spent going from one doctor appointment to the next.

Afterwards the G.I. doctor told us he had something more to offer. Not that this wasn't what I wanted to hear, because it was, but how could it be? All of the other numerous G.I. doctors told us there was nothing more they could do. They'd done every test and offered all they had to give. I always begged doctors to do everything possible.

Next the doctor inquired as to whether we could return in one week's time so my three children could undergo repeat esophagoscopies. Cringing at his request which had totally caught me off guard I found myself clutching my stomach. Once again I had to forget about me and my feelings, and be strong. I had to remember why I was doing this, and do whatever it took. So many things were whirling through my mind. The first thought was, "I haven't even made it through this city yet to our hotel room, which will be a feat in itself, and you want me to come back here next week?" Apprehensive I hesitated for a second, and almost said I'd have to think about it. The next second, I told myself, "This is it. This is exactly what we've been longing for. Another chance and hope!" I replied, "Yes, I can have them back in one week."

Because the children complained of food becoming stuck in their throats while eating, barium swallows were performed as part of the G.I. workups. It had gotten to the point where at each meal, after ingesting only a few bites of food, they would all stop eating due to nausea, stomach pains, or their food lodging in their throats. This was frustrating after I had bought food, prepared meals, and the kids couldn't eat it. Eventually we decided to let them graze feed; eat small amounts throughout the day.

I'll never forget seeing the room in which the barium swallows were being performed fill with doctors watching in amazement as Randi and Brooke showed no peristaltic functioning in their esophagi. Drew's was impaired. (Peristalsis is the muscular wavelike contractions that propel your food through your G.I. tract.) Both Randi and Drew had undergone barium swallows a few years earlier, at which time no abnormalities were found. The initial studies were done, not only because of the feeling of food being stick in their throats and esophagi, but throat pain which was brought on by exertional activities.

One week later, we returned to Johns Hopkins. Directions were still fresh in my mind. All I had to do was a repeat of the previous week and hope for the best. In all of the frenzy, I hadn't had time to stop and think for a second what I would do with three children post anesthesia. One is bad enough, as any parent who has been through this experience can attest.

Brooke and Drew were rapidly discharged after their procedures. They were fine, except they had a difficult time standing up and walking on their own. Unsteady, they swayed and bobbed off each other and everything else around. By myself, I tried to hold onto Brooke and Drew while maneuvering Randi's stroller. She was in a stroller due to her exertional fatigue. Quickly I decided to find the closest bench to allow the anesthesia to wear off a little more. We were not near an entranceway to the hospital where I could have just pulled up and safely loaded them into our van. Generally, and on all other occasions, the children were brought to a hospital exit in a wheelchair. For whatever reason, this time it didn't happen.

Brooke had previously undergone several esophagoscopies, and this was Drew's first. Doctors explained the "something new" they had to offer was staining the biopsies for eosinophils. Eosinophils are a specific type of white blood cell produced in the bone marrow and normally found in the blood and gut lining. Large accumulations and elevations in the body indicate disease. A number of factors can trigger this abnormal immune response, causing tissue damage. Previous doctors could have ordered this simple test.

Randi was unable to undergo her procedure due to an infection and exacerbated asthma, but she had already undergone a dozen esophagoscopies by this stage in her life. I was fortunate to accompany Randi into the operating room on one particular occasion while anesthesia was being administered. While the mask was being placed over Randi's face, she drew in a big deep breath and held it, as if she was going under water... I watched intently, waiting for her to exhale. I was the first to notice, she didn't continue breathing! However, I didn't want to be the first to say anything. I was torn between my instinct to say aloud "SHE'S NOT BREATHING!" or wait for the doctors to notice. I didn't want to be thought of as a parent who overreacted or couldn't handle the experience then possibly be barred from future procedures. Within a minute or so, others noticed.

Someone yelled out, "She's not breathing, she's tightening up."

"Finally." I remember thinking, "Now what?" I was feeling very weak in the knees.

Someone else said, "Is it her reflux? Did she reflux?"

The anesthesiologist began massaging Randi's throat and she started breathing again. Entering the room, just as Randi started breathing again, the G. I. doctor had missed the whole incident.

Post procedure the G.I. doctor stated, "There's something going on in her throat and/or esophagus because it took me a great deal of time and over a dozen attempts just to pass a probe down, which is not typical."

The anesthesiologist came to me after the procedure informing me he had no explanation for what took place. Therefore there was no way of preventing a similar event in the future.

He advised, "Randi should never undergo any anesthesia unless she's at a major medical center where they are equipped to deal with a pediatric emergency."

He said I also needed to let everyone know of the difficulties she experienced with this anesthetic whenever she was to undergo anesthesia in the future.

During our next visit to Johns Hopkins, Randi underwent her esophagoscopy. Because Brooke's biopsy results revealed abscess on top of abscess of eosinophils, we were referred to another G.I. doctor who specialized in eosinophilic disease. I will never forget this date, November 18, 1994, or this doctor! After yet another lengthy appointment, the new gastroenterologist told us he thought he knew what was wrong with my children and there was a treatment available.

This was what we had prayed and searched for for the past twelve years! Was I hearing this man correctly? Could this really be it?

Randi sat beside me in her stroller, the illest of the three children at that time, and said, "Yeah right, I'll believe it when I see it."

My inner excitement was dampened by Randi's words. She was suffering from over a dozen severe, chronic, and debilitating symptoms and problems.

Immediately upon returning to our hotel room I placed a phone call to our pediatrician's office. Although I would follow up with her as soon as we returned home from our medical trip to discuss the findings and plan our next course of action, as was commonplace, I couldn't wait to tell her the good news. After all, we'd waited twelve, long, frustrating years to hear the words I had just heard! No one else was more involved in this fight and journey than our pediatrician.

I thought I was keeping my composure well, but after trying to relay a message to her staff, they asked if they could just leave a message stating, "Mom is ecstatic, please call."

Because of all the medical procedures, we would have to delay Drew's birthday celebration. We have come to accept that medical trips take precedence and dominate our lives. By no means was this the first special event or holiday we were unable to celebrate at home.

We returned home from Johns Hopkins and anxiously awaited Randi's biopsy results from her esophagoscopy. It was highly suspected that Randi's biopsy findings would be similar to Brooke's. All three children had similar symptoms and problems. I questioned the gastroenterologist specifically as to which of the children's multiple problems this diagnosis and treatment would help alleviate.

His reply was, "Hopefully all of the gastrointestinal issues and we'll just have to wait and see if any of the others are related or impacted."

Meanwhile, we finally had an answer. Multiple, complex, protein intolerances was the diagnosis. The children, I was told, could not digest or break down complex proteins in the normal foods they were consuming. Neocate, an amino acid based formula, was formulated to provide proteins in their most elemental form. No digestion was needed for Neocate to be absorbed and used by the body. (Neocate was renamed a few years ago to E.O. 28. However, it will be referred to throughout my book as Neocate. It should also be noted my children were on Neocate One Plus. The One Plus version was for children versus infants, hence the one plus suffix attached to Neocate.)

Neocate was a new treatment. Some insurance companies were not yet covering this special formula. Insurance coverage would be the biggest obstacle facing us and could stand between my children getting better, possibly living, or not. Every new patient would essentially have to fight their insurance company for coverage. Doctors at Johns Hopkins also told us if the children were admitted to the hospital to receive the Neocate, it was covered by the insurance companies. In the hospital setting placement of a nasogastric tube would be required in order to receive the Neocate. A nasogastric tube is a tube placed through the nose and into the stomach. It must be removed and reinserted regularly.

I had always promised myself that the very second I had the answers or help available I would do everything in my power to stop the pain and suffering as quickly as possible! One minute was one minute too long! I finally had the ability to make a difference and help my children! Driving on the long trip home from Johns Hopkins gave me ample time to devise a plan.

Initiating my course of action, we removed all of the complex proteins from the children's diets which we were told are the usual causes of protein intolerances. Besides these, the children already had a long list of aversions or foods known to cause problems. There wasn't much remaining in their diets after this. There was no going back, though; I could not feed them foods which were literally killing them!

Randi's condition was very poor and worsening every day. There wasn't any time to waste. Something had to be done, and quickly! Randi's life was a living hell! She experienced great difficulty falling asleep, and after finally falling asleep, severe pain would awaken her. This was in spite of being on high doses of Elavil (a medication meant to help Randi with the multiple never ending pains which she suffered day and night). Elavil caused her to have significant tremors making it difficult to use her hands. She cried herself to sleep every night and had done so for the past few years. She only fell asleep from pure exhaustion. Throughout the night, she would wake up and grab her throat in what appeared to be an effort to swallow. "The longer she stayed asleep the better," I remember thinking, "because that is less time she will be consciously suffering pain." I even purchased a television for her room, hoping this would distract her mind from the pain.

From the minute she awoke in the morning, Randi experienced severe nausea even prior to attempting to eat. Eating anything only increased it. The act of eating the smallest amount brought on choking episodes which often led to aspirating, wheezing, stomach pains, and at times vomiting, along with a feeling of food being lodged in her throat. She had heartburn from gastroesophageal reflux. For the past few years, Randi had suffered severe migraine type headaches - nonstop. The absolute worst symptom was the constant, intolerable, throat pain which was in addition to the chronic throat infections due to her immune deficiency and dysfunction. She complained of this pain every other minute! Randi's daily numerous loose stools were accompanied by cramping and abdominal pains. Vision and memory loss episodes occurred a dozen times per day.

There were many areas of pain, tingling, and numbness in her extremities. A T.E.N.S. unit, (transcutaneous electrical nerve stimulator), was used to help disguise the pains, but it made Randi look like a robot! She wore the little beeper-sized control unit around her waist, in a little backpack or purse. From this unit, numerous wires connected to electrode pads were placed over Randi's multiple areas of pain. T.E.N.S. was quite helpful, until Randi developed allergic responses to the electrode pads, so this forced us to discontinue using this unit.

(A T.E.N.S. works by sending a signal, felt as a sensation by the user, along the nerve pathways. It's thought the brain can only receive or perceive one signal at a time; therefore the nerve stimulation from the unit blocks the pain signal.)

At eleven years of age, Randi stated, "It would have been better if I died before I was born."

It was a hard thing to hear, but she was aware many genetic disorders end in miscarriage, nature's way of preventing such suffering. She had heard our doctors say this more than once. Randi's favorite number is thirteen because she didn't think she'd live to be thirteen years old.

The constant awful thought was the children's conditions were worsening as they aged and remained undiagnosed and untreated. Although Randi was the worst, Brooke and Drew were following closely in her footsteps. What one developed the others most certain would, in time.

(The phrase 'time will tell' is most applicable with this disorder, unfortunately.) I wondered and feared how much longer they could go on-especially Randi. Family members and others were asking this same question out loud.

CHAPTER 2

TRIALS AND MORE TRIALS

Neocate was our first real ray of hope. Then…our insurance company, Blue Cross and Blue Shield, decided they would play games with my children's lives. Time was of the essence! As much as we hated to, for the children's sake, we informed the insurance company that if they did not agree to cover the children's at-home treatments of Neocate, we were ready and willing to admit all three to the hospital.

After several weeks of correspondences, I gave our health insurance provider twenty-four hours to make a decision. Our bags were packed; Randi, Brooke, and Drew would be admitted to our local hospital to receive Neocate via nasogastric tubes. Due to the children's already significantly damaged esophagi, we didn't want to have to go this route. However, we were out of options. We hoped Blue Shield would come to its senses allowing the children to simply drink the needed treatment at home under non-traumatizing conditions and at a fraction of the cost. It would be ridiculous to admit three children indefinitely to a hospital.

The phone call came within the 24-hour period. A four-week supply of Neocate had been shipped. I was told, along with the formula, we would find important papers to be signed and returned. After receiving the supply and reading the papers, I decided I would not sign. Quickly I realized there was nothing to lose by not signing, but everything to lose by signing them.

The papers, Blue Shield so desperately wanted me to sign and return read as follows: reimbursement for Neocate one plus formula via scientific lab for the four week diagnostic trial period benefits will not be available after the end of the trial period.

Essentially, Blue Shield wanted me to release them from any further obligation in this matter, in exchange for a four-week supply of Neocate. No regard was made as to whether it helped my children or whether they would continue to need it.

Randi's condition had deteriorated to the point where she could no longer swallow her own saliva without great difficulty and without choking and aspirating on it. Even the Neocate we fought so hard to obtain Randi couldn't swallow! All of this happened right around Christmas. We thought, possibly, this newest complication was due to an infection/illness, so we tried to wait it out. On

January 5, 1995, Randi was admitted to Johns Hopkins on an emergency basis under the care of our new gastroenterologist. Further testing was performed on Randi during her admission to the hospital. Randi had always displayed autonomic nervous system dysfunction, (dysautonomia) and the G.I. doctor felt this needed to be addressed.

(Your autonomic nervous system is responsible for controlling involuntary bodily functions such as sweating, digestion, and heart rate automatically.)

Following the tests, Florinef, another medication, was prescribed in an attempt to increase and maintain blood volume and pressure.

Doctors in Boston questioned dysautonomia and Randi was evaluated at an adult hospital there. Unsure of what to make of the results, no form of treatment was ever recommended.

On our way to Johns Hopkins, I could not imagine Randi returning home without some sort of tube required for feedings. Instead, I was simply told, "Just take it slow and keep trying to get the Neocate into Randi." It was hoped the Florinef would make a difference.

To this day I can't tell you why or how, but Randi was slowly able to begin getting the Neocate in and keeping it down. If I hadn't seen slight improvements in her brother and sister, I would have stopped Neocate. I told the doctors, in fact, it was making her worse. More than likely, her overall condition was to blame. This disease had just progressed so far. She had gone undiagnosed and untreated so long. (I learned a valuable lesson about giving something ample time to work. I can't imagine what our lives would be like if we had given up on Neocate.)

It wasn't easy for any of the kids to take Neocate. The formula smelled terrible, like a dirty diaper, and although it was flavored to help hide the taste and smell, it still tasted foul. (The makers have improved it somewhat since, and it seems smoother in consistency and more palatable.) In the beginning, we stirred in separate flavor packets to hide the taste and get Neocate in the kids. All three siblings took quite a while to tolerate Neocate. It was hard to go from foods their bodies rejected, but at least tasted good, to something which was a struggle to ingest.

Randi, Brooke, and Drew's sole nutrition became Neocate. On school days, we packed a big cooler with cups, spoons, flavor packets and Neocate. Roughly every two-and-a-half hours, the kids drank their liquid nutrition.

I recall a few months into Neocate use all three children became ill with a stomach bug that was going around. At first, when I didn't know for sure this was why the kids were ill, I panicked. I wondered, since it was still early on in their treatment, if the progress I saw was all just wishful thinking. Were they really getting better, or were we just hoping they would?

July 1995. It was now several months since starting Neocate. Miraculously, every one of Randi's more than a dozen severe, chronic symptoms and problems had resolved. The children were now experiencing a life free from daily pain and suffering. We watched the symptoms and problems slowly melt away, one by one, just as they had initially appeared. It was a complete turnaround for Randi. Brooke's and Drew's improvements were equally as remarkable and even quicker than Randi's. In Drew, it was approximately three and a half months; and in Brooke, five and a half months. The girls still had chronic infections due to their immune deficiencies but, for the first time in their lives, we were able to distinguish which symptoms and problems were caused by the infections themselves. The girls also were severely visually impaired and learning disabled, but we could deal with that. No longer was there a multitude of other problems which complicated the picture. We finally had something resembling a near normal life. What a change! Prior to this, everything in our lives had evolved around the children's needs and special accommodations made in an attempt to live with this disease.

Because of the children's dramatic responses, which involved far more than just resolution of gastrointestinal issues, there was talk the children might have a metabolic disorder. Suspicion was first raised by someone I spoke with at the facility which developed and manufactured Neocate. This person asked me if I'd ever heard of 'Lorenzo's Oil'. He explained there was a movie documenting the story, and we should rent it. Although we did rent and watch the movie, I felt all of our problems were finally behind us. It didn't matter, why the Neocate had such an impact. We were just so very happy it did! After all, we were accustomed to doctors not having answers for us.

It was very clear, Neocate would continue to be needed. We must pursue a battle with Blue Shield to keep this miraculous, life transforming and sustaining formula. Our pediatrician asked Medicaid to cover the costs, temporarily, while we attempted to figure out how to fight Blue Shield. The girls qualified for Medicaid due to their documented one of-a-kind chromosomal anomaly. However, since Drew wasn't labeled "handicapped," he didn't qualify for Medicaid. The cost of the formula was approximately $600.00 per child, per month. We had already exhausted all equity in our home just to get to this point. There was no money for this added expense. Medicaid stepped in and agreed to help Randi and Brooke temporarily. I could not allow Drew to go untreated. Amongst the three, I rationed Neocate meant for two. At times, we ran short, or completely out, for a few days.

Along with Neocate, we slowly and cautiously began food challenges. The few safe foods the kids had been allowed were bland and had become boring. These were plain white potato, plain white rice, bananas, and plain pasta. Every food item had to be in its plain, simple form. There could be no additives unless those additives had already been tested and were on their safe foods lists. We were educated as to how to conduct food challenges, but chose to make our tests even more stringent. Our food challenges went as follows: on day one, the new food was ingested once. If everything appeared fine, on day two the new food was given three times, morning, noon, and night. If still there were no problems experienced, that food would be given at least once each day for another week. Then, and only then, we considered the food challenge a success, and that food item was added to the safe list for that child. I kept extensive documentation and compiled lists for each child.

A number of foods that were initially not tolerated, during the first few years of food trials, appeared to be tolerable years later. When a child consumed an unsafe food, he or she would display any of a number of symptoms such as migraine headaches, nausea, vomiting, stomach pains, and food becoming stuck in their throat. Sometimes the reactions were instantaneous. Other times, we clearly documented a specific delayed response which was reproducible. The children had to have a perfectly normal baseline before starting a food challenge. If possibly another source was responsible for any symptoms, such as a bacterial or viral infection that may have emerged during the course of the food challenge, then this food would be reintroduced again later.

Besides the very structured and closely monitored food trials we were conducting as a family, Drew had conducted a few of his own. Sometimes these experiments couldn't be concealed. One time, earlier on in his exclusive diet of Neocate, Drew ate a chocolate bar which he was selling to raise money for his little league team. I found him lying in the middle of my bed, all curled up in the fetal position, crying and writhing in pain. I asked, "What did you get into now, Drew?" It was only natural for a ten-year-old boy to want candy. He had to watch as his friends ate candy and ice cream after the games.

Towards the end of Drew's exclusive Neocate diet, we found out that he was taking food from his grandparents' house next door. He would hide in the little wooded strip between our homes and eat it. I didn't keep any foods other than the few that were safe, while Randi, Brooke, and Drew were on their special diets. I ate my meals next door, with my mom and dad, instead of in front of the kids. It was especially hard for them when we went on medical trips, since they couldn't eat what I ate.

I continued to strive to maintain as normal a lifestyle as possible, despite the children's inability to eat. In our society, eating plays a part in most social gatherings. For birthdays, instead of cake, we decorated the top of watermelons and served watermelon. At Halloween, and other occasions when candy was a part of the celebration, the children were given alternative things such as small toys.

At age thirty-five, I felt I could finally return to college, something I had pondered doing for a number of years. Being a single mom I knew it still wouldn't be easy, but now the girls wouldn't need constant daily care. They could return to school, and between their personal teaching assistants at school, whom we trusted, and their grandparents living next door, I thought, "with a little luck it just might work." I had this very strong gut feeling that it was now or never.

I came across an article in the newspaper advertising a medical assisting program. Since the classes began in only a few days, I had to make my decision quickly. In the back of my mind, I knew I had to try to become financially self-sufficient. The children's father had recently left us. There was no telling how

When fighting multi-million-dollar insurance companies, publicity is needed and it is probably one of the best ways to make an impact. Researching a number of articles, I found it was commonplace for insurance companies to deny all requests. A great deal of money and time is needed for an individual to pursue any kind of retaliation. Insurance companies use this to their advantage in dodging their obligations.

If it weren't for Mr. Kevin Luibrand, this would be an entirely different story with a different outcome. I was not deterred by the insurance company. As a mom of three sick children, I was willing and ready to fight! However, this alone would not have been enough to provide my children a life without suffering. Only because of the intervention and help from Kevin was this possible. Forever will our family be indebted to Kevin, his wife, and his firm (Tobin and Dempf), for their efforts and compassion.

If there is a way or means available to stop someone's pain and suffering, cost should never, ever, be a factor or stand in the way of its resolve!

Blue Shield's stance: Neocate was classified as a supplement, and they didn't cover supplements. The big day, which ended up becoming two days, was fought in the New York State Supreme Court. Both our pediatrician, and the children's G.I. specialist appeared and were called to the stand. We did not request our pediatrician to be a witness. Surprisingly, Blue Shield had subpoenaed her. Kevin didn't, because we had our G.I. specialist as a witness. Of course, her presence worked to our advantage, as I was sure it would.

One of the Blue Shield lawyer's first questions to our pediatrician was, " Dr., with regard to the children and their health care problems, this Neocate is no magic formula which is making those health care problems go away, is it?"

Our pediatrician laughed at the question and then replied, "It has made a lot of things go away. It has been amazing the amount of improvement that the kids have seen on this formula. There are other problems that are permanent that are not going to go away, that have nothing to do with this formula."

I remember thinking to myself, "Blue Shield just lost this lawsuit!" I don't know what they were thinking. Then again, it doesn't take a lot of brains to figure out it would have been easier and cheaper to have covered the Neocate to start with and foregone all of this. Blue Shield didn't bargain on our ability to acquire the help we needed to see this denial through!

What close timing! Our pediatrician had been informed by Medicaid only a few days prior to our court date that they would soon stop Neocate coverage.

While I was at work, Kevin called me to give me the good news. The judge had ruled in our favor. He read the entire decision to me over the phone.

I was not familiar with lawsuits since this was the first I had ever been involved in, and hopefully it was the last. Kevin informed us, down the road, there would be further action in the lawsuit. The next round would include a jury. The current court decision would stand for at least a few years until the next trial. When Kevin asked what amount to place in our lawsuit for recovery, I said, "None." I didn't get into this for money.

"You must include an amount," he indicated, so he put the least amount allowable.

My wish and request was for Medicaid and Kevin to be reimbursed for expenses. For years, I received phone calls regularly from other parents who sought help trying to obtain

Neocate. I'd make copies of our decision to send them along with advice on how we obtained coverage. It was hoped that the precedent set by our ruling would serve as a basis for further cases.

CHAPTER 3

THE NEXT BATTLE

We had won our precedent-setting lawsuit with Blue Shield for the children's life sustaining formula. I had recently graduated from college with a 4.0 plus average and high honors. I was hired directly from my internship with a large local doctor's group. The job was part-time, which allowed me to continue caring for my children's needs. I had aspirations of furthering my education and becoming a physician's assistant in the near future. Life was good! The worst truly was behind us now, or so it seemed.

Little did we know while we were away at my graduation, events had already been set into motion that would change our lives again. On college graduation night, my children, my mom, and I went to the ceremony and out to dinner afterwards to celebrate.

Upon arriving home that evening, we found several messages on our answering machine from neighbors warning us of a rabid acting skunk last seen heading towards our yard. We had our three dogs fenced in our backyard. There were no signs of anything out of the ordinary. However, we realized, through earlier incidents with our dogs and wild animals, a situation involving a rabid animal could occur without us ever knowing it happened. We lived on the river, so there were numerous wild animals around nightly.

One night the previous year, I was awakened by a commotion outside in our backyard. Though hindered by the darkness I could tell our three dogs were in a fight with a wild animal. There was blood everywhere, and the dogs were all scratched up. I lay awake half the night worrying I'd find one of the dogs had bled to death by morning. At daylight, to my surprise, all of the blood was licked clean. Days later, one of our dogs dug up a muskrat. That's when I knew it was dead, what it was, and where it had gone. One of our dogs, notorious for digging, (hence his name Digger) had buried the dead animal. If this incident had taken place anywhere else in the yard other than right outside my bedroom window, I never would have suspected such an event could take place.

We didn't want to be paranoid about going into our yard or around our dogs for the rest of our lives. I talked with the kids about wild animals, and dead animals our dogs might find, and what to do in these instances.

It soon became obvious they had listened. As I was working around our pool area one-day pouring cement, Randi let out a blood-curdling scream I couldn't ignore. She was on a swing within sight. When I looked up to ask what was wrong, I saw only the swing swaying.

Randi had quickly slid off the swing and was screaming, "The dogs have that dead animal you were looking for," as she made a mad dash for the door.

It was the skunk our neighbors had tried to warn us about days earlier. It seems 'good old Digger' had buried the animal, probably after all the dogs had mauled it. This meant we had been in contact with our dogs for the past few days since they were exposed. After putting on my gloves, I took the skunk away from the dogs.

I called animal control, based upon our neighbor's information about the skunk acting strangely. The skunk, a nocturnal creature, had been out in the afternoon attacking people and animals. At the time, a neighbor called the sheriff to the scene. She was irate because the sheriff had refused to shoot the skunk. She had tried in vain to prevent what happened to us from occurring.

The animal control officer told me, "Don't worry. It's probably not rabid."

I knew it was!

A few days later, we were notified that the skunk had tested positive for rabies. Our dogs and cat had their rabies vaccines up to date and were revaccinated post exposure. That left us needing rabies vaccinations. Worried and very concerned about the possibility of the girls having trouble with the vaccine, or being unable to build antibodies to fight off a possible rabies infection because of their immune deficiencies and dysfunctions, I contacted our doctors. (Nothing has ever been black and white when it comes to their immune systems.) None of our doctors could answer my questions about these potential problems. Despite the lack of reassurances we had no choice but to go ahead with the vaccinations. The risks were too high not to.

Following the third dose of rabies vaccine, in a series of five, I developed flu like symptoms. Initially, I felt they were probably not connected to the vaccine. I assured myself and the doctors administering the vaccines more than likely I had caught something from the girls or I picked up something at work. After the fourth dose, however, I had little doubt. It was the vaccine. All of the symptoms I experienced following the third dose were now worsening. There were more problems like deep down to the bone aches and pains every minute throughout my extremities. The pain was so severe at times, it made me stop what I was doing, grab the affected area, and rub it. Returning for our fifth and final dose I informed the E.R. doctor at the hospital, I didn't want it and why. He consulted with someone and told me I had no choice. The final dose of this vaccine had to be given.

Well, it was all downhill from then onward. (I've not known normalcy since.)

I began having tremoring in my arms that soon turned into severe chorea, (constant shaking). I was having brain functioning problems such as memory loss, difficulty processing information, slowed responses, and trouble concentrating. I had muscle cramping and weakness, hundreds to thousands of muscle twitches, extreme tiredness, balance and coordination problems, visual problems, increased tendon reflexes, and dystonia.

(Dystonia is abnormal muscle rigidity which causes painful spasms. The sufferer's muscles pull, push, and twist the body into strange movements and/or positions.)

I didn't have a doctor myself because prior to this, I was very healthy, strong, athletic, and energetic. Therefore, one of the doctors I worked for at the time became my doctor. I went from employee to patient in a matter of days.

After briefly examining me, my new doctor replied, "You're f'ed." He said, "You need to see a specialist today."

He made an appointment right away. After seeing several local doctors, being evaluated, and going through tests, which were all negative, I decided to go to Johns Hopkins. My dad drove me to Johns Hopkins, which was chosen based upon my children's previous experience and success with this hospital. However, there were no answers there for me, and all their testing turned out negative. I told myself I'd just have to deal with it, wait, and see what would happen. I was very use to that concept by now. My doctor wanted to admit me to our local hospital, but I refused. There wasn't anything anyone could do for me there I couldn't do for myself at home. I would be far more comfortable and didn't have to leave my children.

For several months, all I could do was lie in bed. I slept at least half of every day and all night. While I was awake, I felt like I might lose consciousness. Other times, I found myself starring into space for what seemed like short periods but in actuality, it was hours. I couldn't focus or concentrate on such things as television or reading material. Any type of stimulus had a negative affect on me increasing and/or worsening many of my symptoms. I became weaker and weaker.

A diagnosis of severe vaccine reaction was the majority of the consensus.

(While educating myself on vaccine reactions, I found anything can happen to you. There's nothing anyone can do to help if you are amongst the few unfortunate ones who have a vaccine reaction.)

I was warned by the C.D.C., Centers for Disease Control, whom I also contacted, not to have vaccines in the future unless there was a life threatening necessity. I was made aware that vaccines administered in a series are most dangerous to a person who has had a vaccine reaction. I remembered the last time I had received a vaccine. It was a tetanus booster, a few years earlier, and I experienced flu like symptoms after that vaccine as well.

Months later, when I was finally able to be up a little, I received a phone call from Mr. Kevin Luibrand. Approximately one year had elapsed since Kevin first began helping us with our lawsuit against Blue Shield. At the close of our lawsuit, Kevin had brought it to my attention that other families were currently waging battles in the Senate and Assembly in New York State for special formulas for their children. We first learned about New York State's formula laws through our pediatrician when we sought to obtain Neocate. The existing law only covered three very specific disorders; none pertained to my children's condition. The Senate and Assembly had expressed interest in our case when it made headlines. I had promised Kevin that when the next legislative session rolled around, I'd assist in the efforts to get a new expanded law passed.

It was now a necessity for us to have such a law. After winning our lawsuit, the children's father, through whom the health insurance policy was held, threatened to quit his job for the sole purpose of the children losing their health insurance coverage. He followed through with his threat, and we were forced to obtain Cobra insurance, which would only last a short time. We'd have to fight, once again, with another insurer to obtain Neocate. A broader law would take care of this issue permanently for us and many others.

Immediately after hanging up the phone with Kevin, I began my mission. It was terribly confusing, for my brain, even to coordinate gathering a piece of paper and pen. I began by calling the contacts Kevin had given me. Those initial contacts were P.K.U. parents.

P.K.U., (phenylketonuria), was one of the three metabolic disorders which already had formula coverage by law. Advocates introduced a new bill a few years earlier, which would provide expanded coverage for some of the low protein foods also required. They needed something to catapult their bill into law. P.K.U., like my children's disorder, if left undiagnosed and untreated, can cause a myriad of problems that can lead to serious complications, chronic and debilitating illness, and death.

The first P.K.U. mom I contacted didn't know me from the man in the moon. She received my phone call out of the blue one day. Initially, I am not sure how much stock she took in my offer to help with this endeavor. I was not the least bit politically inclined. Admittedly, I was clueless about what had to be done to get a law passed. After inquiring as to what needed to be done, I compiled an agenda and began executing it with optimism and confidence. I was on a mission for my children! Anyone who knows me also knows that when I set my mind on doing something I make it happen. 100 percent is what I give to everything I do.

The P.K.U. mom lived near New York City, three to four hours away from Albany, our state capitol. We lived ninety minutes away.

Assemblywoman Elizabeth Connely was the sponsor in the Assembly. Senator George Maziarz was sponsoring a bill in the Senate. My first expedition to the legislature in Albany was to see Assemblywoman Connely. I still had tremoring, weakness, poor coordination, trouble speaking, and my brain was not all there. As I began to try to tell the account of my children's pain and suffering and the need for this bill to become law, I broke down crying.

Fumbling through my speech, I questioned myself, "Do I still have what it takes to advocate for my children?" It scared me.

Next, I put together packets of information pertaining to our cause for every New York Senator, Assemblymen, the Lieutenant Governor, and Governor, which I hand delivered. I felt this was an important step in the process.

A local mom had heard of our plight through a newspaper article. She also had a son with special needs, and was as desperate as we had been prior to my children obtaining Neocate. This mom offered to join our efforts. She and I set up appointments with many of the Senators, Assembly people, or their office staff, and the Lieutenant Governor to tell our stories and discuss face to face our children's needs. We brought these men and women to tears with our stories. No one knowing the facts should hesitate to force insurance companies to pay for these formulas.

Letters were mailed out to all the parents who had contacted me after our lawsuit over Neocate, enlisting their help. Many needed the same coverage we were seeking. Periodically I updated them on the status of our efforts. To keep the ball rolling individuals involved statewide, along with family members and friends, were encouraged to write letters and make phone calls regularly to the legislature. Everyone gathered signatures on petitions throughout the state. A pep rally was held in Albany. During the rally we broke into groups and met with various Senators and Assemblymen. Through our earlier efforts, they were already well aware of our intent. Our gastrointestinal doctor assisted us with writing the actual language of the bill. The day the legislator voted on the bill, my children and I were present in the legislative chambers. Not knowing anything about politics, law or the process, we sat through hours of bewilderment before entering the lobby to ask if our bill had come to a vote.

It had passed!

All that was left was for the governor to sign it. Six months later, it was officially on the books as law in New York State. What an experience! The places we've been. The things we've learned. The people we've met, and the things we've done along the way! Never in my wildest dreams could I have imagined I'd do something like this in my lifetime! I am extremely proud to have been part of such a momentous accomplishment.

The power of just one, caring, determined parent can be truly amazing, but several combined is quite a force! For every law that exists someone had to fight for it. Rule of thumb, it usually takes years of commitment, determination, support, and hard work to accomplish law passage. Contacting and meeting with a constituent is the first step in the process. The Assemblyperson or Senator will work with you to sponsor and write the bill.

CHAPTER 4

LIFE GOES ON

With the law passage experience under my belt, I regained my confidence and felt maybe I could return to work. It was a little over a year now since the rabies vaccine reaction. I was at a semi functional point now, but not one hundred percent. I didn't know if I ever would be again. For financial reasons alone, I needed to give it a try. My assumption was I would continue slowly, but surely, to gain back what I had lost. Trying to get a job locally at doctors' offices or our nearby hospital where they knew of my condition, or of my children's huge medical needs and expenses, stacked the odds against me.

The last time my doctor examined me he said it was unlikely I would ever be able to work in the medical field again.

"Direct patient care is definitely out of the question," he went on to say, "because of how messed up you are physically as well as mentally."

With no luck locally, I applied for a job at a hospital in Albany; a 90 minute drive each way. I got the job, which I really wasn't sure I could handle. I gave my all at work, but I needed to rest in bed almost all weekend for at least the first few months until my strength and stamina improved. For five years, I commuted to Albany and back Monday through Friday. Working the 3-11 PM shifts helped. I didn't have to get up early, and I could still coordinate Randi, Brooke, and Drew's medical care during the day. I loved my job, and the people I worked with. A great deal of pride and accomplishment were felt working in this capacity.

The first five years were mainly as the E.R. secretary. I hated being called a secretary, because I was a medical assistant. It was the only thing I didn't liked about the job. In New York State, a medical assistant cannot perform the same duties in a hospital they are formally trained to do in a doctor's office such as giving injections, medications, and filling out prescriptions. This was fine. I found many new responsibilities and opportunities that weren't available in a doctor's office setting.

There came a point when I reached a plateau in regards to my health and the recovery process. Despite being very active physically, no matter how hard I tried, I couldn't make any further gains. This seemed odd to me, but I accepted it as the way it was going to be and considered myself very lucky to have reached this level.

I had a decent job I loved. I was in control of my children's health insurance for the first time. Things were once again on track and looking up for us.

We planned a rare non-medical trip to Disney World in Florida. (Well, it was supposed to be non-medical. It started out that way).

Both girls ended up becoming ill and requiring medical attention upon our arrival in Florida. Brooke was the first to become ill with an infection. We left prepared with a varied supply of antibiotics, suppositories, inhalers (all our usual items.) When reservations and plans have been made for several months in advance, one makes the best of whatever happens.

Brooke seemed to be responding to the new antibiotics our pediatrician had sent us with, just in case, when Randi fell ill. We enjoyed a few days before Randi reached her limit. That was our last day at Disney. We had just walked into the park when Randi couldn't go on. I sat on a bench, and she laid her head on my lap and cried. I had to come up with a plan quickly. I decided to take Randi to the infirmary there and stay with her. With reservations, I allowed Brooke and Drew to enjoy the park until they were tired, under the condition they remained together.

On our drive home, Brooke's health worsened again. She was unable to eat, drink or take her needed medications. After entering one of the rest stops, Brooke exhibited a strange look. She starred into space and didn't respond to my voice. I knew something was wrong. She began to sway. I got behind her just as she passed out. I gently guided her to the floor. Nobody around seemed to care. They walked around us without even asking if everything was O.K. I guess it looked like we had things under control, and we soon did.

Drew just stood there asking, "What's that fool doing on the floor, Mom?" Typical response from a loving brother.

I then drove eighteen hours straight so I could get her to our pediatrician. It would be very difficult for another doctor who didn't know the girls and their multiple complicated medical problems to care for her. (Obtaining health care is always a major worry when we are away from home).

It was a bittersweet family trip, one none of us will ever forget.

Towards the end of my five years as emergency room secretary, I started noticing a slight increase in the residual neuro-muscular symptoms: muscle twitches, cramping, feeling faint, less energy, and increased weakness. It seemed to be a gradual loss of ground. In the back of my mind remained the unanswered question as to whether or not the vaccine reaction was the sole cause of what had happened to me.

In June 2000, I, along with others at the hospital who had the same health insurance policy, was forced out of coverage. Our premiums had quadrupled and no one, not even the doctors who had the policies, could afford them. The reason stated for the action was that our health insurance claims were too high. (Yes, they can do that and they did.)

I was very upset; no one besides me was willing to challenge this. Health insurance for our family was worth much more than the money I was paid to work. It was of the utmost importance that my children have health insurance. The traditional health policy we held was one of only a very few left in existence which allowed an individual to go out of plan to receive care. Since the majority of my children's doctors were out of state, no other coverage would do. Loss of coverage forced me to start looking for a new job which would offer traditional health insurance as an option. I resigned my full time position and went to per diem status. This allowed me to look for another job while keeping my foot in the door.

Within a matter of only a few days, I ended up face down in bed with a herniated disk in my lower back. I couldn't do much of anything for four months. This wasn't the time to look for a new job.

I went back to Albany Memorial Hospital where I knew I always had a job. There was an opening for a technician on weekends. Working as an E.R. technician was the position I had originally wanted and applied for. For a number of reasons it was a good choice for me. More hands on care was involved, which I liked. I would make as much money in two days as I had previously earned working the whole week. There would be less travel time and expenses and more time home with the kids. Working weekends helped me cope with the increasing lack of energy I was experiencing and my ever-increasing residual neuro-muscular symptoms.

The issue of the children needing health insurance was still present. Consequently, we had to turn to the children's father for health insurance coverage. I was left with none, since we were divorced by this time. He was every bit as cooperative and helpful as in the past. I needed to personally call his employer's human resource department for the information to make a decision about what coverage was available to best meet our children's needs. Then, I set up a court date for the purpose of asking a judge to award the coverage. Their father never showed for the court proceedings, so the judge automatically granted the order. It took weeks for their father to sign the paper work. I wondered how long he would keep this job, and the health insurance coverage. I hated giving up my control over such an important part of my children's lives.

Randi was the only one of the three who hadn't and couldn't wean off Neocate. There were a number of reasons. She wasn't able to find enough safe foods, when we conducted food challenges, to comprise a nutritionally balanced diet. Whenever she tried to decrease her intake of Neocate, her overall health deteriorated, and she became anemic.

Randi's overall condition was so horrible prior to starting Neocate, that being unable to eat most things she took in stride. She knew Neocate was a much better option than going back to the way things used to be. However, to this day, both Randi and Brooke must ingest their formula, Neocate, very slowly or they become nauseated. A time or two, Randi has broken down crying in a restaurant while watching everyone else eat. Sometimes, she is able to have something special prepared when we're out but it is challenging to find safe foods at most restaurants.

Randi's body doesn't tolerate most medicines. She also seemingly has a problem metabolizing medications. When she was younger, despite administering multiple doses of sedatives numerous times, doctors were never able to sedate Randi. Pain medications, if they did anything at all, only took the slightest edge off Randi's pains.

Troubles appeared again in Randi when menstruation began. Although intensified with menses, the problems she encountered were constant. She had hot flashes one minute and chills the next, along with nausea, cramping and diarrhea. What were the odds that she wouldn't have problems? I wished we could have stopped these new problems somehow.

When Randi and Brooke were first diagnosed with their one-of-a kind chromosomal anomaly, doctors predicted that by the time they were old enough to have children, technology would exist to correct their genetic defect. This would eliminate the issue of reproductive problems. Well, that hasn't happened. I would never want to take away the girls' ability to have children in hopes someday they will be cured. I know how important having children has been to me. On the other hand, Randi has expressed, since a young age, she would never want to put a child of hers through what she has experienced.

Next, Randi began developing reactions to her infusions of intravenous immune gamma globulin, (I.V.I.G.G.) Gamma globulin is made by collecting blood from a large number of healthy, normal individuals. These are antibodies normal individuals make in response to infections, which immune deficient patients need and receive via infusion. The I.G.G. undergoes rigorous testing and purification processes to ensure that no diseases are passed to the individual receiving the infusion. However, the protocol is only as good as the technology currently available, and new germs pop up. It can't always be 100 % safe. Twice in the past, we were notified of possible contamination of the girls' I.G.G. The first time was a scare with hepatitis C. Randi and Brooke had antibody testing performed, which was negative. The second time we were informed, one or both had received a batch of I.G.G. donated by an individual who was confirmed to have Creutzfeldt-Jakob disease, the human form of mad cow disease. We were told there was currently no way of knowing whether this particular disease could be transmitted via I.G.G. It would be many years post transmission before symptoms or problems would start to show. Since the girls already had numerous neuro-muscular problems, similar to what one would expect to see with CJD, if suspicion ever arose likely we would never know until postmortem if CJD were to blame. Since the ages of six and five, Randi and Brooke had received intravenous infusions of immune gamma globulin every three to four weeks. It was at that age they were diagnosed with immune deficiencies at Boston Children's Hospital.

By fall of 1999, Randi's reactions to I.G.G. were so significant and the problems they created so severe, we had no choice but to stop the infusions. Brooke also started displaying reactions, but hers weren't as bad as Randi's yet. For a few years, Prednisone was used pre and post administering I.V.I.G.G. to avoid reactions. Early on this strategy seemed to be working. We also employed a slower, less volume, more frequent approach and administered the I.G.G. subcutaneously, into the tissue below the skin, instead of intravenously.

At first, there were either no problems, or they were only mild such as hives and pain at the site of administration. Severe headaches, that developed with infusions or afterwards, eventually became chronic. Nausea post infusion was initially mild, and for only a specific amount of time, but with each subsequent infusion, it became chronic and intolerable. In response to the infusions, areas of severe localized pains developed that lasted several months. No medications alleviated the pains. Injections of Prednisone into the affected areas were not helpful either. Several months after the I.G.G. wore off in both girls, we were finally able to concretely document that the chronic pains, nausea, and headaches had been directly related to the I.G.G.

During Randi's last infusion, in addition to the problems already listed above, she experienced tachycardia (heart palpitations) along with alternating chills and sweating episodes, whole body tremoring, and chest tightness. We didn't know what to think other than (in some way) these symptoms had to be related to the I.G.G. infusion.

That was it! We decided there could be no more I.G.G. infusions. Strangely, symptoms that somehow seemed related to the I.G.G. infusion returned approximately one week after we stopped the infusion. The symptoms would come and go in different combinations, and there seemed to be no correlation as to what precipitated them.

Randi also complained of episodes of feeling frightened, like the feeling you get when you're startled and or scared by something. Randi underwent heart testing and monitoring, but everything except numerous tachycardic episodes appeared normal. An MRI scan of her abdomen ruled out pheochromacytoma, a serious condition which caused similar symptoms. We were then referred to a metabolic specialist, who did a complete blood and urine work up for metabolic disorders. It was all negative. What was going on?

Ninety nine percent of the time that my kids developed new symptoms, all testing would be negative and there were no answers! Only years later, in retrospect, do we find the answers after the kids have endured daily, significant pain and suffering. It's extremely frustrating to know, when new problems arise, it will take years to find solutions. Whenever new symptoms pop-up, we try to get a handle on them as quickly as possible before things become more complicated, and the snowball effect takes place, hindering our chances of finding answers.

We weren't sure what impact being off I.G.G. would have on Randi. She still required high doses of daily antibiotics even while receiving I.G.G in an effort to fight her chronic infections. Because of numerous doctors' concerns with daily antibiotic usage, we had to repeatedly allow Randi to suffer with illnesses to prove she still needed antibiotics. After a local ear, nose, and throat specialist examined Randi for the first time, I was asked if she had ever been evaluated by an immunologist. I laughed and replied, "Yes, numerous top-notch immunologists."

He said Randi presented like a cystic fibrosis patient because of her multiple infections, despite high doses of antibiotics. Eventually he recommended we look into a metabolism problem as the basis of Randi's chronic infections and allergies. Severe, chronic, unresponsive allergies and inflammation were still a constant ever-increasing problem which greatly impacted Randi.

Our pediatrician had attended a conference where a doctor from the Mayo Clinic spoke about systemic eosinophilia. She thought this might explain some of the children's newest unresolved problems since we had documented the children had eosinophilia of their gastrointestinal tracts.

This prompted a three-day medical trip to the Mayo Clinic in Minnesota. It was a very frustrating venture, like so many in the past. Once again, we had to try to convince doctors what we were living and dealing with was real. Part of the extensive work-up performed on the girls for their immune deficiencies and dysfunctions was done in an effort to prove me wrong. The tests only further documented and reinforced many of the controversial issues. The doctor refused to even test my children for systemic eosinophilia, the testing we had come for. I was impressed with Mayo's overall approach, which they are famous for, but still there were no answers and no further help for us.

Our search continued. We began delving into natural, grass roots, non-invasive treatments. After all, this approach carried fewer chances of side effects. The kids were intolerant and allergic to so many things. In dealing with their multiple medical problems, as well as their extensive list of intolerances, we learned we must only conduct one change at a time. Otherwise, we had no way of differentiating what was to blame for improvements or complications. One of the first things we tried with Randi was Probiotics. Unlike antibiotics which destroy beneficial organisms, in addition to infectious bugs, probiotics restore and help maintain a healthy normal environment in the gastrointestinal tract. For the first time in Randi's life, she went from a few weeks to months of being infection free. The majority of the time she was now able to get over infections without developing secondary infections or complications. This, of course, was still in conjunction with daily antibiotic use. From time to time, Randi still experienced long-term infections. Doctors, Randi, and I began to question whether these infections were bacterial. We began exploring the possibility it might be one particular germ Randi's immune system had trouble with, or it might be viral in origin. We would always give Randi the usual ten days to two-week time frame with new infections. Then, if she was clearly not improving or worsening, we would switch her antibiotic in an attempt to gain control of the infection. Unfortunately, when one of these type infections rears its ugly head we know Randi will be out of commission for several months. We continue to investigate the source of these incapacitating, long-term infections. Usually the symptoms consist of a severe headache and sore throat.

When our local E.N.T., (ear, nose and throat) doctor obtained a biopsy of Randi's throat, actinomycosis was found. This is a half fungal / half bacterial infection that can cause serious disease if it's detected anywhere other than in the throat. In an immune deficient patient, it's debatable. Although we had hoped the biopsy results would shed some light on Randi's chronic infections, this wasn't the case.

Every available medication to combat allergies had been tried. Many combinations were used daily with little or no relief. Our family nearly went crazy trying every allergy product and modification known. We purchased drinking water filters and air filtration units for every room in our house. Particleboard and carpets were ripped out of our home along with anything else we felt might be causing or adding to Randi's symptoms. Her whole bedroom was overhauled; only the basics remained. Randi had to give up many of the things that she enjoyed and brought her comfort, such as her stuffed animals. The only thing that was a known allergen we didn't get rid of was the family cat. However, he was not allowed in Randi's room.

Nothing ever worked until Probiotics. They brought another very welcomed improvement to the chronic, severe, seemingly unresponsive allergies and inflammation that Randi had been plagued by her entire life. Randi was able to discontinue all of her allergy and asthma medications after several months of Probiotic use. We didn't think we'd ever see this day! It took us eighteen years to accomplish this.

Brooke weaned completely from Neocate after approximately three and a half years. Through food challenges, we knew several foods she had to avoid. Other than this, she returned to a normal diet. Brooke's senior year, 2001-2002, was when things began to fall apart again for her. This year she mainly spent at home. When she did go to school, a few days a week, it would be for only a few hours at a time. She participated in cheerleading and was in the school play. Often, this was all she could do. This was not your average student situation. The school staff made these modifications because they were very familiar with my girls and their unique complex needs stemming from their chronic medical conditions. Brooke complained of an increased tiredness and weakness. We began pestering Brooke to sit up straight because her posture was changing. She exhibited a rounded forward curve in her upper back, kyphosis. It was clear there was some additional thing wrong with Brooke. She started displaying bizarre behaviors, like staring into space and not connecting with what was going on around her for longer and longer periods.

Brooke always had a problem staying on task. She would mentally drift often and repeatedly. This was not because she had attention deficit disorder. Her mental ability seemed to be declining. Brooke had a widely variable mental capacity, noted by those involved intimately with her education from an early age.

She started having episodes where she remained in the shower for approximately forty-five minutes, and only thought she'd been in for a few minutes. She dumped out medication without being aware of her actions. She walked into the road while working with her orientation and mobility instructor where it was inappropriate and when it would be dangerous to do so. She displayed and complained of confusion episodes. The harder she tried to concentrate, the worse her confusion became. Greater numbers of these bizarre, out of Brooke's ordinary incidents were occurring.

I cautiously mentioned it to our pediatrician (for a few months.) One day I called her and said, "Brooke needs an M. R. I. of her brain. Something major is wrong!"

The results were shocking. Brooke was eighteen years of age, and we were just finding out, for the first time, that she had a malformation in her brain. It was described as hypoplasia of the cerebellum with increased cerebral spinal fluid space making up for the lack of normal brain matter.

When Boston Children's Hospital gave up on us they stated, "We've done every test known to mankind. There's nothing more we can or will do."

Indeed, these kids had endured the gambit of most tests. Randi had a few M.R.I.'s performed at Boston Children's Hospital which were negative. An M.R.I. was never pursued for Brooke because everyone involved just figured, since the girls' problems were so mirrored, there was no reason for one. Doctors and I have since learned we can't always assume this nor should we.

Brooke and I came up with a plan for restarting Neocate. We were quite sure this would decrease or eliminate her new and increasing problems. Taking Neocate or any other substitute form of nutrition, exclusively, is a very hard thing to do. She had to take in enough Neocate to make such an impact. We decided Brooke would eat one nightly meal. She would ingest Neocate for the rest of her nourishment. Neocate would account for approximately seventy-five percent of her daily nutritional needs, 1,250 of her daily calories. The one meal would consist of any of her safe foods.

This seemed to, once again, return Brooke towards her baseline, but a two hour nap was still necessary every day.

Drew weaned from Neocate completely after two-and-a-half years. He only needed to avoid a few things in his diet. He was the first of the three to discontinue Neocate and the one who was least affected prior to its use. This may have been because he was the youngest, and he didn't have immune deficiencies complicating matters.

Although on the surface things seemed fine after Drew quit Neocate, they weren't for long. As time went on, Drew shared with me his constant fatigue which caused him to nap most days after school. I also knew he once again had chronic headaches.

Drew would state, "There's something wrong with my metabolism," during checkups with our doctors. However, he couldn't and didn't elaborate.

None of our regular doctors had any further ideas other than recommend we go to Europe for care. Unfortunately, this was off limits financially, so natural medicine doctors were consulted. Electrolyte solutions, Magnesium, and Colostrum helped alleviate some of Drew's symptoms. This new approach, now employed with all three children, was very expensive. We couldn't afford to keep all of the prescribed treatments going. These medications were not covered by insurance. Environmental / Natural health doctors concluded my children had an inborn error of metabolism. The problem was that we had yet to find a mainstream doctor who could confirm, document, and treat this so that it would be covered by insurance.

I was not aware of how serious things had become for Drew. Being a sixteen year old male, it was understandable for him to keep medical problems a secret until he no longer could.

One day I received a phone call from the school. It was Drew's senior year. I was told I needed to pick up Drew because he had been suspended for the possession of pills in his pocket. I couldn't imagine, as I drove to school, what this was about.

When I walked into the room where Drew sat, he blurted out, "Mom, can we please talk about this at home?"

Of course, I wanted to know what was going on right then and there.

Drew then said, "My whole body is all messed up Mom. I have aches and pains in all my muscles."

I knew what this meant. I was very upset not because Drew had pills in his possession at school, but because Drew's condition was deteriorating again. The pills were Daypro, a prescription pain/anti-inflammatory medication. Drew had taken them from his grandmother's medicine chest in an attempt to alleviate the multiple areas of pain he was experiencing. The pains in his hands were so intense at times he could barely hold the trumpet he loved to play. At other times, he had difficulty with muscle cramping in his throat so he couldn't blow his trumpet.

I wasted no time. I brought Drew to see our pediatrician that afternoon. She was able to document the whole body tremoring he experienced. He weighed approximately twenty pounds less, which was also noticeable. To her he described the pain he was dealing with and the nauseousness which was worst in the morning. For years, he had been complaining to doctors about significant fatigue and headaches.

Due to Brooke's array of new neurological problems, our gastroenterologist referred us to a colleague of his in the Department of Neurology at St. Christopher's Hospital for children.

After examining Drew, ordering thyroid studies, and other basic blood work, our pediatrician said she would call this latest neurologist to consult with him if this current testing came back negative. Our pediatrician was quite sure, however, that Drew's new problems were straightforward and due to a thyroid problem.

I said, "Wouldn't that be nice if it were something that easy." Not to my surprise, all the testing was negative.

Our pediatrician called me after speaking with the neurologist who said he knew what was wrong with Drew, Brooke, and Randi. She told me he felt our whole family was afflicted with mitochondrial disease.

I remember replying, "Ya think so!" I was just venting my bottled frustrations.

For some time I had arrived at the conclusion our family shared some sort of a rare disease. I based my theory on a number of indicators. First, there were the children's past and present histories of multiple unexplainable symptoms and problems; then, my own similar struggles with multiple neuro-muscular symptoms and problems, which were also unexplainable and worsening as time went on. Several months prior to my severe vaccine reaction, I experienced many of these neuro-muscular symptoms. At the time, I began taking a complete multi-vitamin and mineral supplement, hoping this would do the trick. Now looking back, I realize this disease process had already subtly begun.

(It's a known fact that sometimes vaccines can exacerbate and/or accelerate a disease process that was dormant but would have activated later.)

My mother and my maternal grandmother had several unexplainable symptoms and problems I felt were, more than likely, related. I encouraged them to share my children's diagnosis with their doctors. This was useless, and meaningless to their doctors who, like the majority, had never heard of a mitochondrial disorder. I also thought this possibly explained what doctors were frantically searching for when my brother was dying of an unrelated cause several years earlier. Doctors felt he had some significant underlying disease or disorder which complicated and accelerated his condition. Our specialists, at that time, were contacted and consulted. However, we didn't know then what we know now.

I had resigned myself to dealing with my problems on my own. I was discouraged by the certainty that if doctors couldn't find the answers for my children, they weren't going to find them for me. All I could expect was to be told I was crazy or depressed. This analysis was one I didn't need on top of everything else. I knew eventually one of two things would happen: we would someday finally find a doctor who would have the answers and a diagnosis; or one of us would die from this. I let everyone know that if something happened to me, I was not to be resuscitated and I wanted to donate my body to medical research for the sole purpose of finding answers to help my family.

Our most recent neurologist had never seen Drew. Randi was present during Brooke's initial evaluation with the neurologist, and I made a mental note of how it seemed he was unofficially examining Randi as much as he was Brooke. I felt good about his perceptiveness.

I had always dreamed of someday walking into a doctor's office and having the doctor tell us, "I know what's wrong with the children," but by this time, I had given up this hope. That was exactly what was happening, in essence, and I wasn't even fully aware of it.

Until Drew's involvement, we hadn't discussed the mitochondrial diagnosis to any degree. It was clear this doctor spoke our language. This neurologist fully understood every one of the children's numerous problems. In fact, everything that were concerns or issues for us were things he discussed. It was evident as well as impressive; he knew specifically what to look for. This alone assured me the diagnosis he made was correct.

Initially, I held back my own problems, for the most part, or at least the extent of them. I needed to make sure the diagnosis was for real. I didn't want my problems to interfere with my children's care. I had been so traumatized in the past by getting my hopes up time and time again with every new doctor and every test that was performed. It was an extremely emotional roller coaster ride. It took a few years of follow-up appointments with the children's neurologist before I relaxed and let my guard down. Nightmares of returning for follow up visits and having him say, "I am sorry, but I don't think this is the root of your family's problems after all," plagued me.

While working through accepting answers after nineteen years, I told myself and the rest of our doctors we could finally relax and switch gears. I thought we could leave the daily search mode and now deal with the disease. What little relief I managed to enjoy, and had anticipated would be permanent, soon dissolved.

CHAPTER 5

WE HAVE THE DIAGNOSIS. NOW WHAT?

Everything pointed to mitochondrial disease, but I wanted desperately to have any available testing done to confirm this diagnosis as soon as possible! All of my children were, once again, starting to deteriorate before my eyes. My worst nightmare had always been the children's problems might return someday. I needed to know if this truly was the answer! It wouldn't do us any good to be misdiagnosed again. If we assumed this was it, and it wasn't, we'd lose more precious time as the children's conditions declined. To my dismay, this confirmation would be an arduous feat in itself.

We were furnished with the names of two specialists in the country who could perform a muscle biopsy and had the facilities available to run the very specialized testing that was required. They were also the only two, we were told, who would accept insurance as payment. The first doctor's office we contacted informed us the doctor was not accepting new patients. They would call us in several months "if and when" he was. The other physician wasn't much more promising. There was no way I could have all three of my children seen, declared the person on the other end of the phone. His office staff took my children's information and said they'd forward it to the doctor. The doctor would decide "if and when" he would see any of my children and get back to us.

It took months of repeated phone calls to get this doctor to respond. Only after conveying to the doctor's secretary I was leaving the next day to make the ten hour trip to speak with the doctor did he call. I intended to camp out in his office until he met with me. On my way out the door to go to work that Sunday afternoon, I received a call from the doctor. He would only consider performing a muscle biopsy on Drew. Since Randi and Brooke also had a one-of-a-kind chromosomal anomaly, we might not know, if something was found, from which disorder it stemmed. I understood this reasoning. At least it was something.

Didn't these doctors care? Didn't they understand? I didn't want my children's health and well-being to slip backwards! I couldn't let that happen! Everyday that went by, I feared we were losing more and more ground and the chances we had of holding onto a better life were drifting away.

My children's lives, as young adults, were beginning to take shape. They were all seniors in high school. We were trying to plan for their future and their health would dictate what they were able to do.

We traveled to Cleveland, Ohio to have Drew evaluated. This doctor wouldn't consider any other family history in making his decision. He was so rude. He insisted Randi and Brooke leave the examination room, not because he was exposing Drew, but because he didn't want his decision swayed in any way by the girls. At the end of our visit, he said he'd get back to us. He believed it was unlikely anything would be found from a muscle biopsy. This was just his opinion, however. I asked him to please think about this from my perspective. I explained that a number of times in the past testing which wasn't initially performed later provided valuable answers. What if I hadn't pushed onward? Likely, my children wouldn't be alive. This doctor brought to mind numerous other doctors we'd dealt with far too often. It was maddening to say the least!

As we were driving back to our hotel room, I was expressing with Randi, Brooke, and Drew my dislike of this doctor and his opinions.

Drew commented, "I agree with him. He's the specialist. He knows what he's talking about. If he doesn't think he'll find anything in the muscle biopsy, why should I have it?"

That was it! I'd had it! After all, who did Drew think I was doing all of this for? I screamed, "You don't understand all of what I've gone through to get to this point. I would never put you through this if I didn't think it was necessary and important. I'd rip a piece of muscle right out of my leg, right now, and give it to the doctors, if I could. I wish I could do this for all of you."

Being so upset, when we returned to the hotel room I had to take a walk. Never before had I left my children alone, especially in a strange, unfamiliar place, but Drew was more than capable of watching and caring for his sisters until I returned, if I ever did.

This doctor had four reasons for hesitating to do the biopsy; first, the actual chance of finding anything was slim. Only a percentage of the mitochondia are abnormal, which usually dictates to what degree an individual will be affected. During the muscle biopsy, it's impossible to know which mitochondia are normal and which are not. Individuals sometimes go through several muscle biopsies before defective mitochondria are found. Second, even when faulty mitochondria are discovered, no treatment options are available beyond those already being used. Third, there's always the risk associated with anesthesia when performing the muscle biopsy. It should be noted; not all muscle biopsies are performed under general anesthesia. It all depends on the center and their protocol. Fourth, doctors also know, even with multiple muscle biopsies being negative, this doesn't always rule out mitochondrial disease.

Evidence of mitochondrial disease can be detected in blood, brain tissue, and /or spinal fluid as well. There's also a non-invasive approach some specialists feel yields parallel results of a muscle biopsy. This is called M.R. Spectroscopy, an application of the standard M.R.I., which measures the bio-chemical make-up in the brain and/or muscle tissue. Still, Muscle biopsies are considered the golden standard. Other times mitochondrial diseases can only be found after a family member dies and extensive testing takes place postmortem. In these cases, it can take months to years for a concrete diagnosis.

Upon returning from my walk, I had a discussion with Drew. I would not force him to undergo a muscle biopsy if he really didn't want to have it done. I reiterated I was only doing this out of love and devotion to help him as well as his sisters. It was a gamble and he was the only one of the three who could help us find the answers we had searched a lifetime for.

Although the doctor remained reluctant, Drew was willing to go ahead with the muscle biopsy. On June 5, 2002, Drew and I returned to Cleveland for the procedure. The results took approximately three months to process and obtain.

On our first attempt, the odds weren't in our favor. The muscle biopsy was negative. We had held off starting medications until after we received these results. In September of 2002, all the children started on Creatine. This was the first of many supplements we would use in an effort to combat some of the symptoms associated with mitochondrial disease. Creatine is mainly used by athletes and body builders to increase muscle mass and strength.

I was barely hanging in there with my own health. I mentioned to my children's neurologist the extent to which I was presently afflicted, and he told me to treat myself accordingly with Creatine. We had learned through the years to care for ourselves the best we could. This was nothing new. For several months, it had taken me all day Monday, Tuesday, and part of Wednesday to recover from working the weekend shift. I had to rest in bed. I felt so frustrated. I hated feeling like this. This wasn't me; this was the total opposite of the person I once was. In the past, when forced to sit still and be inactive it would nearly drive me crazy. Now, basically it was all I could do. My earlier suspicions were confirmed; the vaccine alone was not to blame. I had already fought this monster once with every ounce of my being. I knew if it ever returned I was in big trouble. The second time around, I knew what to expect, which usually makes things a little easier. However, I didn't want to live the rest of my life like this. It didn't take me long to realize some people, including my own children, have never known normalcy and may never. I was ashamed of myself for complaining. Fortunately, I had experienced good health for the first thirty-five years of my life and had envisioned being one of those people still very active and going strong at ninety years old. This disorder, however, is stronger than the strongest person.

The Creatine and Neocate I was taking helped some, but not enough. By February, 2003, I had met my match. This disorder had gained the upper hand and has dominated my life ever since. I was knocked to the ground, and I couldn't get back up. At times, I was unable to do even the simplest of things. Even my baseline was no longer normal. For no reason, I experienced shortness of breath. Just breathing caused chest pain which was present most mornings and would often awaken me in the middle of the night. During the day, my chest also began hurting. The only things to relieve it were lying down and resting, or a short-term fix using nitro tablets. I concluded it had to be muscular in origin, because it seemed, at times, to solely involve my chest muscles; other times it affected my upper back muscles or a combination of both. Sometimes, it seemed due to weakness or tiring of the muscles. At other times, it was more muscle cramping and dystonic in nature. This symptom alone was present at least half of a 24-hour period and stopped me from doing most everything. I was struck by episodes of profound weakness out of the blue, forcing me to lie down. At their worst, I couldn't move a muscle or talk. This was occasionally accompanied by nausea. The one thing I could compare these weakness episodes with was the way one feels when stricken with the severest case of the flu, (feeling as if you can't do anything, and you just don't care). I was starting to have dystonic episodes after years without them. More and more often food eaten at dinner wasn't going anywhere. I was waking during the night choking, vomiting, and/or having diarrhea. This would last for hours leaving me very drained. Massaging my stomach for a few hours alleviated my discomfort. However, I didn't want to keep reliving this. It took me a while to figure out, but I eventually concluded this was dysmotility, the very same thing my children all experienced. My gastrointestinal problems are manageable if I eat dinner before three or four in the afternoon and consume smaller meals. At least several times a day I felt faint and was having severe, chronic, right-sided headaches which originated behind my eye. At times the headaches didn't respond to the maximum dose of a combination of three painkillers. I noticed a decline in my brain functions. While at work, I feared inadvertently causing harm to a patient because of this. Most of the time I caught my mistakes.

Well aware of what it was like from a patient's point of view, I couldn't continue working in this capacity. I took my job very seriously.

Unless I get to the point where I can no longer function, I will not go to a doctor. This was one of those times. I wanted to investigate the possibility I might be dealing with something which could be treated, but I was without insurance and unable to afford medical care. My only option was our local hometown health center since I couldn't pay for services. It took several weeks to be seen by a doctor. First, a P.A. had to officially indicate I needed to see a doctor, something I was already sure of. The P.A. also confirmed my need to go on disability, something I was certain of but didn't want to accept. The physician agreed to order basic blood work which would rule out anemia and other treatable maladies. I explained to her what I assumed was the root of my problems, the same thing that affected my children.

She replied, "That's not an issue. I am an internal specialist. I can handle the case."

She promised, two or three times before she left the room that she would get back to me with my blood work results, and we'd go from there. Well, after waiting and waiting for this to happen, I placed a phone call which was never returned. I finally called back and asked for copies of my blood work. This was how I found the testing was normal. When I attempted to make a follow-up appointment with this doctor, I was told the next available opening was in three months.

What was I to do? I couldn't work to the capacity needed, but I kept on trying to do what I could. Things were not getting any better, only worse. I didn't want to mix work with my needs but I was desperate. When I finally got up the nerve to ask one of the doctors I worked with for help her only response was, "You need to see a specialist."

Because I didn't have insurance, I was denied even an x-ray of my chest to rule out anything obvious that might explain the chest pain which prohibited me from functioning. This was a hard pill for me to swallow. I worked in a position dedicated to caring and helping others, but when asking for help I really needed, it was refused. Every weekend trying to work reinforced my need to seek help. I couldn't go on this way.

Once again at our local health center, with another doctor, I explained everything.

She replied, "First, I've never met anyone who's had a vaccine reaction, and secondly, I don't know what a mitochondrial disorder is. I've never heard of it. Is it something new?" Then she said, "A lot of your symptoms are somewhat similar to chronic fatigue and fibromyalgia. I could diagnose you with that and treat you accordingly."

I felt my anger boiling up inside. I replied; "Why would you do a stupid thing like that? Forget it! I will just have to wait and see the specialist."

An appointment had been set up with the specialist my children's neurologist had referred me to for evaluation of mitochondrial disease. I waited far longer than I should have to ask him for the referral, because I had never faced anything physically I couldn't overcome before. Then, it was seven months before I could be seen.

One last attempt was made to obtain help, or at least some answers, for myself by going to our local hospital emergency room. I knew if I walked in complaining of chest pain they would perform the chest pain protocol. Previously, I had an E.K.G. and repeated them on myself at work. It wasn't my heart. I already knew this. After spending the entire day in the E.R., the doctor I saw stated he wasn't aware of any doctor in our area that could or would take on such a case, and in turn referred me back to my hometown health center. I was so frustrated!

Days before I was to leave for Pennsylvania to see the specialist, I received a call from that office saying they had to reschedule my appointment. I was crying and pleading , "You can't do this to me, please don't do this!" I told them how I had barely been hanging in there for the past several months.

A conference had come up the doctor must attend, and there was nothing they could do. I would have to wait six more weeks.

I met with my nurse manager at work and explained what was going on. I said, "I will do my best. I may need to take time off from work, leave early, and I can no longer work more than eight-hour shifts." I also asked to only be assigned the desk position versus the technician position. Sitting wasn't a much better option; it was a very hard thing for me due to the chest pain. Because I had earned everyone's respect over the years, I was never questioned about my failing abilities. I only divulged my health problems to a few close friends at work.

At last, I saw the specialist in November of 2003. She ordered hundreds of tests, all of which I was familiar with because my children had undergone them in the past. This was the first step. The tests were to rule out any other possible reasons for my symptoms. She placed me on disability as well, but it wasn't as easy as the doctor saying she's putting you on disability. Aware of the hassles facing me down the road when I had to apply for permanent, long term, social security disability, I hated the thought, but I didn't have any other choices. I feared being unable to financially care for my family; we were barely making ends meet the way things were.

Lacking insurance, I had to apply for Medicaid and have the testing performed in Albany, N.Y. Months passed again before I could see a specialist in Albany who was supposedly familiar with mitochondrial diseases. My initial appointment was scheduled for 6 A.M. the day after Christmas. My past medical history, along with my children's, was mailed to the doctor with a letter letting him know why I was seeing him.

He started by being smart and saying, "So, how did you get to Pennsylvania?"

I replied, "I drove!"

I was not in the mood! It had taken nearly a year just to finally begin the testing. I reiterated what I had already made very clear in my letter to him. Only because I had Medicaid did I need to have this doctor.

He started yelling at me and said, "Well, if you don't want my fine medical expertise that's O.K."

Then he left the room. When he returned he sort of apologized. We were both clear now, at least, on the only reason I was there. From the report he generated, based on the one office visit, it was obvious to me he was like the other hundreds of doctors we'd seen in the past. He would never have diagnosed my children or me with a mitochondrial disease. My children and I had all been seen in Albany by numerous specialists on a number of occasions. This is why we were receiving our medical care in Pennsylvania!

From years of experiences with my children I had learned to always obtain copies of test results! It's a good thing I did. Although I was told everything was negative when I received the results, I found approximately one third of the testing wasn't done. Nobody but me seemed to care either. If I hadn't followed up, more time would have been wasted. I had the doctor reorder the tests that were not done. The second time around, all but one test was performed. It was my serum lactate level. (This is one of the important tests when dealing with mitochondrial diseases.) I made the three hour round trip to Albany a third time specifically to have a lactate level done. I explained to the laboratory technicians the special handling and manner in which the specimen had to be processed. This was the third attempt, to get the the testing completed and I would give them an hour to allow them time to prepare. Despite all of my efforts, my lactate level was never performed in Albany and I was never given an explanation as to why. I wasn't about to waste anymore of my time, money or energy on this endeavor in Albany.

Usually Lactic acid levels can only be properly processed at major medical centers which are familiar with handling these specialized samples. Although Albany Medical Center Hospital is often touted as being a major medical center, capable of things found at other major medical centers, our family's numerous experiences there have always proven Albany is not on the same playing field.

After everything else had finally been ruled out, my physician in Philadelphia referred me to another specialist in Atlanta, Georgia for a muscle biopsy. This doctor could perform a very detailed, extensive, mitochondrial work-up, along with the actual muscle biopsy. After all these years I finally had the chance to do something which could help my children. The only way I could come to grips with the return of my problems was to think of it as a way of finding further answers and help for my children. Documentation was now an important issue for me as well so I could receive social security disability, (S.S.D.) My temporary disability insurance would only last six months, and I hoped I would have my biopsy results before S.S.D. made its decision.

I poured what little energy I had into trying to persuade Medicaid to approve coverage for my muscle biopsy. The twenty some years of daily searching had come down to this. I called on my usual arsenal for help. I asked all of our doctors to write letters of support for us.

Our pediatrician stated in her letter, "After all the hundreds of thousands of dollars of Medicaid funds that have been spent on evaluation and treatments for the three Evans children, it makes the most sense to get the best information available from this biopsy. Inaccurate or incomplete testing of Mrs. LaFond-Evans' muscle tissue will be quite useless".

To deny the testing which could finally bring us the answers was ridiculous! I wrote approximately fifty correspondences over a several month period to the director of Medicaid's prior approval unit. He admitted he had no knowledge of mitochondrial diseases and promptly denied me the out-of-state muscle biopsy. I was not asking this for myself, but rather for my children.

We consulted with a number of lawyers and the consensus was, "You can't fight the state."

We were already several years into two other existing battles with Medicaid. The first was an attempt to obtain coverage for the multiple supplements needed by the children costing between three hundred and five hundred dollars per month. The second was my daughter Brooke's extensive dental needs which had gone unaddressed and unmet. I found out Medicaid's "so called" appeals process was bogus; in fact it didn't exist. We never received a reply from the several correspondences sent via certified mail regarding these matters. The phone numbers provided didn't yield any more success. I documented and had other individuals document that repeated phone calls were put on hold for hours until we hung up.

Another area of dispute among the specialists who deal with mitochondrial diseases is whether fresh or frozen tissue is needed for the biopsy.

When Medicaid did their research into muscle biopsies for mitochondrial diseases, they were advised I only needed a muscle biopsy performed on frozen tissue. This was in contrast to the fresh tissue our doctors wanted for more specific testing. Originally, in one of Medicaid's correspondences, they stated there were two facilities in New York State that could do a fresh tissue biopsy.

In actuality, no facility in New York State could perform a biopsy on fresh muscle. Only one facility in New York State could perform the frozen tissue muscle biopsy, Columbia University in New York City. Medicaid gave me the name and number of a contact doctor at Columbia University who in turn recommended a doctor to perform the biopsy. It was months of near daily phone calls to this doctor's office just to find out he didn't accept Medicaid, and that was why my phone calls weren't being returned. I was furious! Medicaid not only forced me to have my biopsy performed in a facility that couldn't test for everything our doctors wanted done, but the doctor they referred me to wouldn't even accept Medicaid. What a joke!

I decided to give up my Medicaid benefits. This was the only way I could have the muscle biopsy performed at Columbia University. I would have to borrow over two thousand dollars to have a frozen tissue biopsy performed, just for a stab at answers.

In New York City hospitals, if you are a Medicaid recipient, you must receive your care in what's referred to as a Medicaid clinic. You are segregated from people who have insurance. If you have a rare disorder, resident doctors in the Medicaid Clinics can't and won't help you. Our out-of-state hospitals don't treat their Medicaid recipients in this fashion.

I was advised by the very doctor Medicaid referred me to at Columbia University, "I wouldn't recommend a biopsy done in a

Medicaid clinic. It would be no better than having your muscle biopsy done in Albany."

I was desperate for any chance, no matter how slim it was. Thank goodness, the twenty thousand dollars which is customary for the actual testing of the muscle itself couldn't be charged because at Columbia it was considered research.

Medicaid never paid one cent. I returned my benefit card with a letter stating, I'd rather die than ask Medicaid for anything for myself ever again. We also canceled Medicaid-funded programs for my daughters, with the exception of medical based services. I question why, if Medicaid can't provide for a person's most basic health care needs, do they fund other things. I don't blame doctors for not wanting to deal with Medicaid!

This was not our first run-in with such nonsense. Several months earlier, we were referred to a major New York City hospital that accepted Medicaid, Brooke's only source of dental coverage. Our pediatrician and I felt they would be knowledgeable about rare disorders and would be equipped to deal with a medical emergency if one arose while Brooke was having dental services performed. Brooke was only allowed to be seen in a Medicaid Clinic. It wasn't staffed by doctors that could handle Brooke's complicated case. After two medical trips, we were told not to return.

I received a phone call the afternoon before my muscle biopsy was to be performed. It was the doctor's office letting me know my procedure might be postponed as the doctor was called for possible jury duty. Why does this always happen to me? Fortunately, my biopsy didn't have to be rescheduled. It was just one more unwanted nerve-racking situation. This added stress; the need to wake early and drive a long distance along with the need to fast for the procedure brought out the worst of my symptoms for this doctor to see. My whole body was tremoring rigorously by the end of the procedure. A friend who had accompanied me to drive home after the procedure didn't know what to think when she saw me after the biopsy. The doctor had decided to use my upper arm muscle versus my upper leg, the usual muscle biopsy site. My arm was in a sling and I was tremoring uncontrollably from head to toe due to exertional fatigue. Surprised by my appearance, my friend questioned what happened.

She said "I thought you fell and broke your arm and now you're in shock, I knew I shouldn't have left you."

She's fortunate to have returned! During my muscle biopsy, my friend, whom I refer to as a luddite, left the hospital to find something to eat and rest. My ride home fell asleep on a park bench in the middle of New York City all alone and awoke to find three large men standing over her holding shovels and rakes.

Our second attempt to document our family's mitochondrial disease was also a strike out. All the turmoil was in vain because everything that should have been done wasn't done.

The one valuable thing we were able to document was my elevated lactate level. I finally had this performed while at Columbia U. the day of my muscle biopsy. This was important useful information.

My family's entire future hung on these results. My initial S.S.D. claim had been denied, which I understood was commonplace. The lawyer I had hired for an appeal felt I needed positive documentation of mitochondrial disease in order to receive S.S.D. It had been several months since my temporary state disability benefits had expired.

Going into the hearing, I made sure my multiple symptoms were obvious. The whole week prior to the date of the hearing, I stressed my body physically as much as possible. It was pathetic what minute exertion it actually took to bring out the worst of this condition. When I arrived at the hearing that morning, I was stressed mentally! Neither my doctor nor my children's specialist had submitted the letters they promised to my lawyer. They were the only doctors involved and knowledgeable on our family's cases. I couldn't believe it! Right then and there, I put in calls to their offices. How could they do this to me, to our family, with so much at stake? They were supposed to help us; they both knew the time frame in which we had to have this information! This really put me over the edge.

As I walked into the room, the hearing officer said, "Now don't worry. We're just going to talk about your case."

How could I not be worried? I was sick with worry and had been since the very day I had acknowledged the fact I had no other option but to go on disability. The decision he made today would determine our being able to continue living, financially speaking. Our family had had no money coming in for several months, with the exception of Randi and Brooke's modest S.S.I checks. All bills and expenses had been placed on credit cards, hoping a favorable decision would be made.

One of the girls' teachers, who became an advocate for our family, went with me to the hearing. She was able to relate what this disorder had done to our family through the seventeen years she'd known us and just how real it was.

Fortunately, my hearing officer took into account the impact the lack of insurance had on my case and our efforts to produce answers. He also seemed to be a caring, knowledgeable person. I was informed within approximately one week that I was eligible for S.S.D. and would start receiving benefits in a few months. I don't want to have to ever go through that again! My fears were well-founded. I could have ended up with a hearing officer who had no knowledge of mitochondrial disease, like all the doctors in the past.

There had to be more testing available. I still wanted everything that could be done to be done. The children's neurologist / mitochondrial specialist performed an ischemic forearm stress test on Drew. Although Drew had a normal lactate baseline, post activity his lactate levels were seventy percent higher than normal. This test measures how the muscles utilize components of cellular metabolism during exertion. However, the results didn't fit into any of the known patterns of abnormalities seen before. The findings only raised more questions. It was a piece of the puzzle, nonetheless.

Ironically, a neurologist at B.CH.H. had performed this same testing on Randi and Brooke just prior to the Boston doctors turning their backs on us. We were told they didn't know what to make of the results. For them, it was easier to ignore the findings rather than investigate what they actually meant.

This former neurologist from Boston called unexpectedly one day to ask me if Randi had a stroke yet. I was very puzzled by this question that came out of nowhere. I asked, "What are you talking about?"

He explained that the short visual and memory loss episodes Randi was experiencing were thought to be mini strokes. They felt, but had never discussed any of this with us, that Randi would probably have a large-scale stroke.

At every follow-up visit with my children's neurologist / mitochondrial specialist, I'd inquire about the M.R.Spectroscopy studies. This was seemingly, our last hope for the concrete proof I desperately wanted of mitochondrial disease. It took our doctor years to find a location where the research was available.

In August 2005, the girls and I went to Columbia University in New York City to have the studies performed. Drew was working and couldn't take the time off. The doctor to whom I had previously been referred for the purpose of performing my partial muscle biopsy at Columbia University had also recommended I follow up with M.R. Spectroscopy testing. He felt sure I had a mitochondrial disease. However, he wasn't even aware the M.R. Spectroscopy study was available at Columbia University. We were told the results would be completed and sent to us in approximately two to three weeks. It turned out to be a five month wait.

My children's doctor didn't fight to obtain the reports; I had to. He didn't seem to care if we ever received them. He was convinced from day one, when he first met our family, we had a mitochondrial disease and the test results themselves wouldn't change anything. He conveyed to us that the researchers who performed the studies, at no charge, may not want to share the results.

In January 2006, after months of phone calls and letter writing demanding the results, we finally obtained them and our third attempt was a strike! A big one! However, we had to wait a few more months to discuss the findings with the children's neurologist / mitochondrial specialist. After twenty-three years of searching and all we'd been through, we still had to wait just to be able to ask the question, "Is this it finally?"

When I did have the chance to speak my mind at our next appointment with him, I bluntly asked him if he'd ever heard of a telephone! I would think he would have known just how important this little piece of information was to me. Apparently, doctors have a far different perspective on obtaining and notifying patients of test results than parents do.

We now had the documentation we'd sought my children's entire lives! The results confirmed the respiratory chain in the mitochondria was compromised. The respiratory chain is one of the very complex processes that takes place within the mitochondria. Unfortunately, it doesn't pinpoint where the exact defect in the respiratory chain is, or tell us what form of mitochondrial disease we have.

The M.R. Spectroscopy findings were only demonstrated in Brooke. Of course, I questioned this. I was told, just as the muscle biopsies are a hit or miss, so is M.R. Spectroscopy. It all depends what the chemical make-up of the cell is at the exact moment of film capture. I was very relieved to have, once and for all, confirmed that our family was affected by mitochondrial disease. I realized it changed nothing from a treatment standpoint. We would have to continue our daily battles the best we could. Nonetheless, it was closure for me! It validated me as a person. I had finally proven all we had gone through from day one was for a purpose, and all of it was REAL!

CHAPTER 6

THE COMPLEXITIES AND CHRONICITY OF IT ALL

Since the diagnosis of mitochondrial disease, a lot more things make sense. However, despite that more and more pieces of the puzzle are coming together, many mysteries remain. Mitochondrial disease is very complex, chronic and often debilitating and deteriorative.

Overall the children had seemed somewhat improved on Creatine, but in time the benefits became stagnant. Drew couldn't tolerate the Creatine in the form we could afford . Randi discontinued its use feeling it no longer made a difference for her. After stopping and restarting Creatine, it was established Brooke definitely still needed it and benefited from its use. Carnitor, another medication in the long list of typical mitochondrial drugs, was not tolerated by Randi or Brooke. Drew has felt some further improvement with Carnitor and CoQ 10. All three have low levels of Co enzyme Q10. Only Drew remains on CoQ replacement, because he was the only one to note an appreciable difference. This does not mean the girls shouldn't also continue its use but at this time, the cost is prohibitive. Mito medications in combination are referred to as the mito cocktail. There are conflicting opinions among doctors as to whether or not an affected individual should continue taking the mito cocktail regardless if improvement is noted or not. You can see how each person's response is different to the various medications, even when an entire family is affected by the same disease.

I have compiled a chart listing my family's symptoms and problems to date. Please, keep in mind there are many forms of mitochondrial disease. I wish to show that even when an entire family or a number of individuals in the same family are all affected, they will have some similarities, but they will also all be affected in different ways. What a person reading this chart needs to understand is from minute to minute, the effects change. An individual can experience any number of, or all of, the symptoms which are unique to that individual, at any given time with little explanation. When the individual does experience any one of the symptoms, it will last for varying amounts of time and to a varying degree of severity. Stresses of any type, whether physical, mental, or both, have a negative impact. Examples of stresses are illness, injury, menses, exertion, cold and heat to name a few. Some of these factors we have control over sometimes. I've found that sticking to what is a tolerated amount of physical or mental exertion is the best way of minimizing exertional fatigue along with the multitude of problems and the severity in general. You have to pace yourself. If you over-exert, any number of your symptoms can intensify. Your body begins shutting down, and you are forced to stop functioning all together. When my children were younger, there was no explanation for this phenomena witnessed in them. Now, only after experiencing this first hand, I understand. My children never had the normal amount of energy other kids displayed, especially the girls. There were times in the past when I pushed my children's limitations before understanding this part of the disease. Of course, now I feel badly about this. It's one of those things you must experience to fully understand. Just because you can do something one minute without difficulty doesn't mean you will be able to the next minute. This makes it impossible to commit to anything.

Mitochondrial disease is a constant fight waged against your own body and brain 24/7.

We once had an unknowledgeable E.R. doctor ask my daughter, "So what do you do with yourself?"

I wanted to strangle him! Instead, I nicely replied, " We fight this disease every minute of every day!"

Mornings are rough for all of us in different ways. It takes our bodies hours to "regulate" after awakening before beginning to function at our baselines. It's a major struggle every morning knowing we'll have to endure and overcome hours of unpleasant symptoms before we can start our day.

Sometimes, even family members don't understand. There are comments made about just being lazy. Sometimes the words aren't said but the eyes roll and the look is given to imply it. My own son, who is also affected, is ashamed of me because I no longer work. He tells me I am capable of doing more than I do. This is very hurtful to me. He bases this on how he is affected and to the degree he currently is afflicted. He's still able to maintain the upper hand. I tell him I wish he understood and hope he will someday, but I don't want him to have to find out the hard way. He can't comprehend how he could suffer his whole lifetime with this and all of a sudden his mom who was always perfectly normal ends up worse than he in such a short period of time.

That's exactly what can occur in families sometimes. There is so much variability and lack of predictability with a mitochondrial disease in anyone who is affected. There's no way of calculating if he'll stay at his current level or whether he will also continue to decline, and to what degree. The uncertainty of it all is one of the hardest parts of dealing with this disease. I don't have a lazy bone in my body!

At twenty-one Drew doesn't understand many issues. I am still giving one hundred percent or more every day just to function to the extent I am. It's extremely hard continuing to prove your self-worth when the very means enabling you to express it are diminished. Drew also is not, and never was, as affected as his sisters. This isn't saying he doesn't have a lot to deal with. He's just starting out in life, and like all young adults his age, is facing a myriad of normal challenges. On top of this, he faces the personal struggles of trying to live with mitochondrial disease. I confided in my son how I had to come to grips with not being able to do what I wanted, when I wanted, the way I wanted. It would have eaten me alive if I hadn't. I am willing to give up my life in the hopes further answers and help could be found for Drew and his sisters. However, the girls would be left with no one that could care for their multiple needs. For this reason alone, this is not an option.

Then there's my daughter Brooke. Of the two girls, Brooke is able to go out into the community the most. Although she is quite limited in what she can do, we find things to keep her happy. There are times when I am driving Brooke home from an activity, (I like to refer to it as driving Miss Daisy), when we're both either near or beyond our physical and mental tolerable limits. We both look at each other and just laugh because we recognize we're in the same boat.

I find myself relying on my daughters to help me now, especially since Drew is away at college. Neither Randi, Brooke, nor I may be capable of everything all the time. Together, with our combined efforts and input, we're able to handle most things that come our way.

OUR FAMILY'S SYMPTOMS / PROBLEMS LIST

Problems	Randi	Brooke	Drew	Mother
Hypoplasia Cerebellum		X		
Optic Nerve Defect	X	X	mild	
Ptosis	X	intermittent		
Strabismus	X			X
Visual Impairment	X	X		
Brain Function Involved	X	X	X	X
Learning Disabilities	X	X		
Seizures		X		
Confusion	X	X		X
Irritability	X	X	X	
Memory Loss	X	X		X
Headaches	X	X	X	X
Dizziness	X	X		X
Vertigo	X			X
Coordination problems	X	X		X
Balance Problems	X	X		X
Hearing Problems	C.A.P	C.A.P		X
Ringing in Ears	X			X
Autonomic Nervous System Dysfunction	X	X	X	X
Low Blood Pressure	X	X		X
Tachycardia	X			X

Problems	Randi	Brooke	Drew	Mother
Palpitations	X			X
Chills and sweating	X			X
Mottling	X	X		
Tremor	X	X	X	X
Blue/Purple lips, nails	X	X		
Syncope		X		X
Cold Intolerance				X
Muscle twitches, cramping, spasms, weakness, and pain	X	X	X	X
Hypotonia	X	X		intermittent
Dystonia				X
Fatigues easily/Exertional fatigue	X	X	X	X
Lax Ligaments	X	X		
Joint Pain	X	X	X	X
Tendonitis	X	X		X
Neuropathy/Neuralgia	X	X		X
Gastrointestinal peristalisis problems	X	X	X	X
Gastroesophageal reflux	X	X	X	X
Food/Protein Intolerances	X	X	X	
Eosinophilia G.I. Tract	X	X	X	
Malabsorption	X	X	X	
Immune Deficiencies	X	X		

Problems	Randi	Brooke	Drew	Mother
No Frontal Sinus	X			
Allergies	X	X	X	mild
Asthma	X	mild	X	
Eczema	X	X		X
Vaccine Reaction				X
Heart Murmur			X	
Digit Malformed	X			
Urinary Problems	X	X	X	
Dysmenorrhea	X	X		
Bone Density Loss		X		
Disc Dessication	X	X		
Kyphosis		X		
Other Back/Spine Problems	X	X		X
Shortness of Breath	X			X
Decalcification Teeth		X	X	
Stiffness			X	X
Weight Loss 10 - 30 lbs.	X		X	X
Lactic Acid Elevation			With Stress Test	Mild Baseline

In the midst of trying to obtain proof of our family's mitochondrial disease, an array of new problems arose for Randi, Brooke and Drew.

With approximately two years of almost non-existent allergy symptoms, for the first time in Randi's life, her allergies returned and were every bit as severe as previously if not worse. I examined every possibility for this turnaround but found nothing that explained it. Our home environment alone is not to blame because Randi's allergy symptoms are greatly increased by going outdoors or other places. Whenever we go on medical trips and stay in hotel rooms, Randi's allergies go haywire. The intensified symptoms take days, sometimes weeks, to calm down after we return home. Randi's life is controlled by her extreme allergies which are unresponsive to any treatment. When doctors or nurses ask what Randi is allergic to, I reply, "the world."

She is allergic to chlorine in pool water. Pool therapy was one of the few tolerable, beneficial forms of exercise for Randi. I bought an expensive copper ion water purifier which would allow Randi to enjoy our pool. However, just going outside causes Randi to become sick to the point where she is incapacitated.

Therapeutic horseback riding was another excellent source of activity, but Randi was highly allergic to the horses. Nine times out of ten when Randi leaves our house, she becomes sicker. If I had the resources, I would build Randi a safe wing on our home. It would consist of only natural materials. There would be Hepa air filtration throughout. A portion would be all glass. Randi could have sunlight and enjoy the outdoors without being exposed to elements of the environment that cause her to become ill.

Presently we are dealing with Randi's newest intolerance, latex in her underwear waistband. She can no longer wear underwear. Where will the intolerances and allergies end? What's next? Does everyone in our family have to stop wearing underwear? Are we supposed to post a notice on our front door advising those who enter to remove their underwear prior to entering our home? Yes, we're aware of the alternative non-latex elastic-banded underwear available at fifteen dollars a pair! I just ordered five pairs for seventy-five dollars. Just one more expensive item we can't afford but need.

Both Randi and Brooke started having anaphylaxis type reactions with unsafe foods. We now carry epi-pens in our arsenal of things not to leave home without. The reactions start like a true anaphylaxis attack. Instantaneously, the girls develop tightness with trouble swallowing and breathing. If they happen to be in the act of swallowing, the second this occurs, they choke and often aspirate. After the choking hazard has been addressed and passes, I sit with them, epi-pen in hand, and wait until the symptoms start wearing off. Generally after half an hour the difficulty with swallowing and breathing wanes. Sometimes the symptoms aren't totally resolved by bedtime. With true anaphylaxis, the symptoms and problems will continue to escalate unless treated. Doctors and I theorize that because the girls have a poor functioning immune system, they aren't capable of mounting a full-blown anaphylaxis attack. This is one time I am glad they have immune deficiencies.

Randi began having increased neuropathy pains in her extremities along with increased injuries to her extremities. Both girls have poor coordination, generalized overall weakness, lax ligaments, and hypotonia. Despite not doing very much physically, they injure themselves easily and repeatedly. It always takes a very long time for the injuries to resolve. The girls regularly have multiple areas of injuries and pains.

Each week when they go to physical therapy, the first question from their therapist is, "So what hurts today, and how are last week's areas of pain?"

Randi actually awakens injured simply because she slept with her arms, hands, or feet in an awkward position. Our neurologist theorizes Randi doesn't receive the signal to move off the affected area.

A number of E.M.G.'s, (electromyography, a test used to measure muscle and nerve function), documented Randi had demyelinating lesions at her elbows from an early age. For a few years, after starting Neocate, Randi seemed to actually display improvement with her neuropathies. To everyone's surprise, we were able to demonstrate remyelinization through two follow-up E.M.G. studies over a three to four-year period. Myelin is a protective sheath that surrounds the nerve fibers; therefore demyelinization refers to loss of myelin and remyelinization refers to the regeneration of myelin. The most recent E.M.G. results indicate the symptoms may stem from the central nervous system versus the peripheral nervous system. We don't know for sure. Doctors tell us if small nerve fibers are involved, which is likely, this can't be documented by E.M.G. studies. Randi suffers from a lot of back pain as well. She has had to return to wearing ankle braces due to lack of control of her ankles, a recurring problem for her over the years.

Brooke has had to wear a knee brace since a young age due to similar instability.

Randi has developed peptic ulcer disease within the last year. With her second flare-up, we wondered if the peptic ulcer disease was secondary to a long term viral infection we had documented, herpes simplex virus type one. Our G.I. doctor had no problem placing Randi on one more life-long medication to take care of the peptic ulcer disease. I would rather get to the root of the problem, if possible, and treat it, because the girls are currently taking well over a dozen medications daily. Each medication brings with it the potential to cause more problems. The number of medications the girls consume is deceiving because we really do limit and refuse a number of medications doctors recommend. If a medication doesn't offer any noticeable improvement, it's discontinued.

It was felt that an upper endoscopy with biopsies would tell us whether Randi's gastrointestinal problems stemmed from peptic ulcer disease or were due to a systemic viral infection. Randi and I spoke at length separately, face to face, with each of the doctors involved, about the need for specific testing in order to confirm our suspicions. I was not going to put Randi through an esophagoscopy under anesthesia unless it was absolutely necessary. Having just gone through a messed-up biopsy with Brooke, Randi and I reiterated the ordeal and concerns to our doctors. After our discussions, I felt sure things would go smoothly and we wouldn't have to worry.

Desire and reality are two separate entities, unfortunately. Situations similar to our recent experience have occurred time and time again over the past twenty-three years since the birth of my oldest. This particular ordeal is still fresh. So, I have a great deal of unresolved emotions attached. Others in the past are now reduced to a mere sentence of overview. In the scheme of things, someday this latest mess will also be. My patience has worn very thin. Immeasurable frustration created by these situations makes me want to tell all of our doctors to go to hell! I wish we had no need for further dealings with doctors. I'd simply like to tell them, "Give us a call when the time comes you can do something to help us." That would be the easy way out.

After I've had some time to cool off and think things through, I have picked myself back up and fought the fight that must be fought. In retrospect, I know if we didn't keep going every time a doctor told us they didn't know, or gave up on us, or accused me of something, we wouldn't be where we are today. I know my children are better off because of perserverance despite the added overwhelming frustrations.

On the day of Randi's esophagoscopy, the G.I. doctor told Randi and me that he and our immunologist had a long conversation as planned. The immunologist requested only the usual biopsy testing performed. I just held my head in my hands and said, "No, no, we are here specifically to rule out a viral cause to Randi's peptic ulcer disease and determine if the current long term infection is systemic." I knew that very specialized, specific testing was needed to obtain the results we were looking for. As

Randi lay there on the gurney being prepped for the operating room, I wanted to say, "Let's just leave." We had invested a fair amount of time, effort, and money into this medical trip and were putting Randi through a potentially dangerous procedure. I allowed the whole process to continue only because I still felt there was a slim chance something might be found. I felt stupid being backed down from our planned course of action. How could the immunologist have said this? It was contrary to the conversation I had with him. Maybe I misunderstood the immunologist. On the other hand, I trusted our G.I. doctor with whom we had a very long good relationship.

A month later when I spoke with both our immunologist and G.I. doctor regarding the biopsy and testing that was done, compared to what we had originally discussed performing, I received two conflicting stories. I didn't know who to believe, and I don't know where the discrepancies stemmed from. I am enraged that either of them, fully aware of the circumstances, would allow this to happen. I'll be the first to tell anyone not to trust doctors but sometimes you have to. These doctors were both long standing and knowledgeable on my children's cases. We did trust both of them. That's the hardest and most hurtful part of this. All Randi went through and we have no further answers.

The one thing I do know for sure is that all doctors, for a number of reasons, are stressed beyond their capacity. They don't have the time to do what needs to be done as thoroughly as it should be.

I feel I am always viewed as the bad guy in these situations. I am only advocating for my children, driven passionately by my desire to rid them of pain and suffering. I want to know why it's so wrong to want to find answers and potential help! If this were their child, it would be a different matter. I acknowledge my children's cases are, and have been referred to by many as, "A big mess!" I hear repeatedly it's just part of it all...part of the whole disease process. This preconceived mind-set clouds and hampers our efforts. Going into any test or procedure, the odds are against us even if everything goes the way it should.

In our doctors' defense, if a viral or multiple viral's were found, the treatment options available would remain the same. Randi had already begun high doses of the herpes medication, Valtrex. HSV is the only viral, with the exception of A.I.D.S., for which there is a medication. It was felt this medication alone couldn't rid Randi of the infection, even if HSV was the sole viral. Randi would require I.G.G. in combination with the herpes medication. I.G.G. was not an option due to the well-documented reactions and symptoms it caused in Randi.

Our specialists told of an experience they had with another mito child. This child also had an infection with an atypical response to I.V. I.G.G. Based upon this, our specialists theorize that what we think is a *reaction* to I.G.G. may instead be the germs mounting a *response against* I.G.G. The other patient had to be admitted to the I.C.U. and placed on life support while a higher than usual maintenance dose of I.G.G. was administered. The I.G.G. was then able to eliminate the infection which previously caused an improper response against the I.G.G. itself. Following the ordeal, this child could tolerate I.G.G. infusions and benefit from its long-term use by helping keep him free of infections.

Neither Randi nor I feel this is something we want to chance to prove or disprove why Randi has reactions to I.G.G. unless Randi's infection becomes life threatening. Granted, I hate watching Randi suffer daily, for months on end, unable to help her.

We must somehow decide which is the lesser of the two evils. I have reservations because there are too many other unexplained factors that may be related which must be taken into consideration.

We decided to discontinue Valtrex use since we didn't think it was helping Randi and we couldn't use it in conjunction with I.G.G. She still had the constant, significant headaches, sore throat and oral lesions. Her gastrointestinal symptoms had resolved after starting Prevacid. Randi's nausea returned and worsened every day until we restarted Valtrex. Seemingly, this would suggest the H.S.V. infection did exist in Randi's gastrointestinal tract after all.

Without I.G.G., the girls' immune deficiencies are not treatable at this point in time. We must face this fact: doctors cannot help the girls with this complicated aspect of their mitochondrial disease. We must continue to deal with much of what comes our way on our own. The majority of doctors we encounter still don't understand or know how to deal with or treat immune deficient patients. I worry if the girls contract a life-threatening infection that it may be too late by the time anyone figures out what should be done.

When Randi was very sick with her documented viral infection, I was being assured that she couldn't have anything too serious. Local doctors unfamiliar with Randi were basing this on the following: her ESR rate wasn't elevated. Randi has never had an elevated ESR rate in her life. ESR stands for erythrocyte sedimentation rate. It is a basic blood test routinely done to measure the body's response to certain illnesses. When an individual's ESR is elevated, it may indicate the presence of infection, or it may be abnormal because of other factors. Then I was told Randi's white blood cell count was normal. The white blood cell count is another routine lab test that, when elevated, can indicate the presence of certain types of infection. If her white blood cell count was high enough to be within the normal range, this alone should confirm illness in Randi and send up red flags. Randi has never had a normal white blood cell count in her life. Next, I was told all of her viral titers were normal. The fact Randi has been documented as not being able to produce antibodies to a number of germs makes this form of testing unreliable and questionable.

In addition to atypical lab findings, when Randi and Brooke present with acute infections, their examinations do not reveal the usual recognizable signs. Standard gauges, such as fever and lymph node enlargement, used to measure the severity of infection are either very subtle or nonexistent in the girls. Unfortunately, the girls are every bit as sick as a person who displays the usual signs of infection. It could be dangerous if they aren't treated appropriately and in a timely fashion.

By 2003, Brooke had reached the point where she, too, could no longer tolerate her I.V. I.G.G. The girls had been evaluated by multiple top-notch immunologists, but all have said I.G.G. intolerance is extremely rare, especially when all the protocols used to address the issues have been employed and exhausted. We have been told over and over again, by every immunologist, there is nothing more they can do to help with this issue. They tell us we must choose between the complications which develop due to their immune deficiencies and those that develop from the I.G.G. infusions themselves. Neither is an acceptable choice! Both girls have less than half the amount of normal I.G.G. Brooke also has no detectable I.G.A. now.

A few years earlier Randi was documented, for the first time, as having lost about half her amount of I.G.M. (I.G.A. and I.G.M. are other immune globulins found in individuals with a normal immune system that also play a role in fighting infection.)

Brooke returned to daily antibiotic use after a year long stretch of chronic urinary tract infections while she was still on I.V. I.G.G. Brooke has done very well, in regard to her chronic infections, on daily antibiotics alone. She still gets more than the normal number of infections: on average, one a month. She is able to get over them in a normal amount of time and there are rarely complications.

In the fall of 2004, Brooke began complaining of increased weakness and a feeling like she had something she couldn't get over. There were no obvious signs of infection. After several months of this complaint, we switched her antibiotic which had no impact.

The spring of 2005 brought about the beginning of even more unexplainable problems. They included a red painful rash with vesicles on Brooke's hands. This was brought on by extended periods of sun exposure. Then Brooke had a return of severe, chronic, erosive esophagitis. None of the kids had experienced this

since beginning Neocate over eleven years ago. Brooke's

constant, at times, intolerable lower back pain had increased to the

point were she could barely walk and needed to lie in bed for days

at a time. Repeat M.R.I. films documented desiccation of every

disk evaluated. Next, she had a return of the urinary symptoms she

had experienced at an early age but had no trouble with since age

nine. Then, there was documentation of bone density loss. This

seemed to go hand in hand with the disk desiccation and tooth

decalcification. Why was Brooke having all the signs of

malabsorption? Around the time we were documenting these

findings in Brooke, both Randi and Drew also stated they felt they

were not getting any nutritional value from foods. About this time,

Brooke began having extensive dental problems and needs. This

couldn't be a black and white problem either.

Although Brooke's dental options were, and continue to be

extremely limited because of Medicaid, the dentists who have

examined Brooke lacked an explanation for the intense tooth and

gum pains she's had for over four years. She requires constant

high doses of pain medication to deal with this alone.

It was fall 2002 when Drew went off to college, something I had hoped this disease wouldn't prevent him from going. Like a lot of moms, I gave his room a thorough cleaning after he left. What I found was more proof of just how badly Drew was suffering. I found a daily journal from his tenth or eleventh grade English class. Every page in it started the same: "I feel sick again this morning, and I don't know why."

My thought was, "If I hadn't been working, would I have noticed?" I wasn't awake in the mornings when he left for school, since I didn't get home from work until after midnight. I left for work before he got home from school. If I only knew, I would have tried to help him by involving doctors.

Drew is still bothered by significant physical and mental symptoms that are markedly worse first thing in the morning. He states his brain feels like mass chaos upon awakening. He has a very hard time functioning, mentally and physically, each morning. This is evidenced in his display of intense irritability, brought on by the subtlest noises. Medical trips are very stressful with all of us cooped up in the same car and then the same room. This is one of the main reasons Drew now opts out of most medical trips. When he does go along, by morning he is always sporting toilet paper swirlies protruding from his ears and is transformed into a totally different person from his usual calm, polite, gentle, easygoing personality. Drew's significant irritability has always been a major symptom, one that has been hush-hush until we met the children's neurologist / mitochondrial specialist. It was one of the first improvements noted by Drew after beginning Neocate. Our pediatrician would say we shouldn't mention this to other doctors for fear they'd think we were crazy. This also meant we didn't address the problem, which wasn't right either. An individual's needs are not, and can't be, met when a doctor doesn't understand or have knowledge of a disease.

My son, much to my displeasure, took up smoking cigarettes. He found the nicotine helpful in calming his tremors and irritability. I don't like it but do feel and agree when doctors can't and won't help, you need to do whatever it takes to get some relief.

By summer of 2005, Drew had his own lawn maintenance business which enabled him to work at his own pace around his illness. Even with this flexible time schedule he was having difficulty functioning.

He came to me one-day and said, "Mom, I basically totally crashed. It's as if I am not getting minerals, vitamins, or any nutritional value from foods."

Because I had forced him onto the exclusive diet when he was younger, Drew had resisted restarting Neocate. He told me he hated me for doing that to him. It was the fact that it was exclusive he didn't like. I told him I did it because I love him. I couldn't take the chance he would become as affected as his sisters. Drew began Neocate consumption again!

CHAPTER 7

PRECIOUS GIFTS FROM GOD

I married when I was nineteen. This is what I wanted. I had hoped to have children at a young age. I wanted them to all be close in age. Then, I could get on with my life, or so I had wished. I don't think I could have handled what I have if I hadn't been young.

My mom and dad married young and started a family right away. My two brothers, sister and I were close in age. I had a great family growing up and I wanted the same. I was working as a machinist when I went on maternity leave for my first child. My pregnancy was relatively uncomplicated. The worst part was the morning sickness I experienced every day of my pregnancy. It was not only in the morning; I was sick all day and on a medication called Bendectin to help relieve this. My husband and I took Lamaze classes and, although the instructor told the class odds were that one mom in the class would end up needing a Cesarean section, I was sure it wasn't going to be me. I was young, athletic, and very healthy; no way. I felt sorry for the person who ended being the statistic!

On May 26,1982, at approximately 11 P.M., my water broke. I had stayed up late to watch a movie that night. I was bored, tired of being pregnant, and overdue. Just as I planned, I drove myself to the hospital. Everything was under control and I was ready!

As time progressed, so did my contractions. Unfortunately, I didn't dilate at all. My contractions were excruciating and close, but still no dilation. I was experiencing the worst pain of my life. I was finally given a cervical block along with Pitocin, a medication to speed up delivery. The cervical block was a welcome relief, as it blocked all my pain. However, it was only temporary. I knew if they were giving medication to induce labor further, when the block wore off and I still wasn't dilated, the screaming they had heard so far would not compare to the screaming I would be doing. I am no wimp. I was a tomboy, and not only could I dish it out, I could take it. As the cervical block began wearing thin, I recall the nurses placing a call to the doctor. I was rushed into the operating room and within minutes, I was prepped for a Cesarean section. At that point, I could have cared less if I had to have a Cesarean. All I knew was that I wanted this pain to stop!

I was fully awake. I had received an epidural which causes numbness only in the desired area. I was told I had a baby girl. She was not breathing so they began resuscitative measures as I watched intently. Soon she began to breathe on her own. Unconsciously, I had been holding my breath until she began breathing. It was now twelve noon. Twelve long hours of labor and then a C-section. I was young and very tough, but I fell asleep shortly thereafter and did not awaken until my family awoke me later that evening. I wanted to see my little girl as I had only seen her briefly at her birth before they whisked her away.

Our pediatricians said she was perfectly normal. Before we ever left the hospital, I had my doubts about that. Randi never opened her eyelids. I was assured by one of our pediatricians in the group that there was nothing wrong with my child's eyelids or eyes. I was also assured the nurses had seen her open her eyelids. This set me off on the wrong foot with this doctor from the start.

Little did I know, nor could I ever have imagined, this was just the tip of the iceberg with regard to Randi's medical problems. At one of Randi's well-baby checkups, months later, the same pediatrician finally acknowledged there was something wrong with Randi's eyelids. He told me she needed to see a specialist because she had a rare condition called ptosis (droopy eyelids.)

"All right, we can live with this," I remember thinking.

We were referred to a specialist in Albany. I was terrified to have to go to the city, one and a half hours away. I didn't want to. I figured we'd go this once, find out what the doctor had to say, and that would be it. Yeah right, that was just the beginning. I've come a long way from that person who dreaded going to Albany. I wonder if I would have ever left Warrensburg, New York, our hometown where my kids and I were raised. My children have broadened my horizons in many ways. My life, although not changed for pleasant reasons, changed dramatically for the better.

Within a few months after birth, Randi began having chronic infections. The infections just lingered and took forever to resolve. As soon as she was over one, or sometimes even before, she had another infection. I figured somehow this had to be normal. Randi was my first. I knew nothing else. Randi began daily doses of prophylactic antibiotics between infections. She has been on daily antibiotics since that time. Initially we attributed all of Randi's developmental delays to her to ptosis. Randi had strabismus (crossed eyes) as well, but so did I. She had surgery to correct the strabismus at age one and a half. The corrective surgery had to be postponed several times due to her chronic infections. At the age of two and a half, Randi had bilateral leviator surgery performed. Both eyelids were cut and the inner muscles pulled and stitched tight. This was only partially successful. Not impressed with doctors' ideas to try performing further corrective surgery, we felt it better to wait. We were told the next surgery would permanently open Randi's eyelids, but then she could never shut them, not even to sleep. She would require daily eye drops, her corneas could dry out, and then she could become totally blind. All of this risk was solely for cosmetic reasons. I chose not to pursue any further surgery for her eyelids. I decided to leave this decision up to Randi when she was old enough to make such a decision. I don't regret that decision at all.

As time went on, eye doctors also noted Randi had optic nerve defects, the extent and impact of which could not be fully documented through vision testing until Randi was older. These defects caused further visual impairment for Randi. She had no peripheral vision, only tunnel vision.

Randi was thirteen months old when her sister Brooke was born. Brooke appeared normal and was dubbed normal by our pediatricians. Approximately twenty-four hours after her birth, Brooke was found in the nursery, blue, and not breathing. She was resuscitated and given to me afterwards to feed. I was told she merely choked on her own mucus and that this happens frequently. To comfort me, and as a precautionary measure, Brooke was placed overnight in the special care nursery for observation and feedings. Brooke was nothing you could call normal, by a long shot, from this point onward! Our first night upon arriving home from the hospital she became very congested, as she had been after needing resuscitation. Concerned, I phoned our pediatrician's office. It was after hours, so one of our pediatricians called from home. Although he apologized later for his actions, he yelled at me for calling and assured me Brooke would live. This was the same doctor who had been sure there was nothing wrong with Randi's eyelids or eyes. This gave me very little comfort! I was quickly finding I couldn't trust doctors. My instincts and gut feelings were right before, no matter what anyone else had said! Brooke never slept or ate well throughout her infancy. The one thing I will never forget is Brooke's almost constant screaming! We were living at my parents' home then. We had extra help, but no matter what anyone tried, it didn't stop Brooke's screaming.

Brooke was admitted to our local hospital a few times her first year of life because of her irritability. Other than an infection and gastroesophageal reflux being documented, nothing was ever found. Nurses told doctors that Brooke was just stubborn. It was written in Brooke's chart that her admission, on at least one occasion, was for parental respite. I was told by our doctors not to stay with Brooke at the hospital. I was still far too trusting of doctors at this stage. Brooke, like Randi, had chronic infections. Still, I felt this had to be normal somehow. After all, they were both the same. Family members kept saying this was not normal, however. Brooke as well as Randi was delayed in motor skills, fine and gross. They both started walking around sixteen months and Brooke never crawled.

My son, Drew, was born sixteen months after Brooke. This was it. We had our family. I had truly wanted to have more, but we had to draw the line somewhere, so we decided this was a good place. Those three C-sections were rough but worth it. I could get my body back to normal, enjoy my children, and watch them grow and play together. They would always be very close, I hoped, since they were so close in age. I could never have imagined just how close and how much they would indeed all have in common! I was once again assured by our pediatricians before ever leaving the hospital with Drew that he was normal.

"He is a placid baby in comparison to his sister Brooke," I was told.

Drew's problems, early on in life, were subtle and compared to his sisters not even considered problems. At birth, Drew had a small skin tag on his ear, the shape and size of a pea. Jaundice developed a few days post birth, but these were the only things that appeared out of the ordinary. A few months later, hydrocele of his testes was noted, but this, too, was no reason for alarm or concern.

He didn't seem to grow at a normal rate. As an infant, we had even discussed performing a bone growth study on him because of this but it was never done. He had unexplained diarrhea often. Drew developed whole body rashes. It looked worse than it was, we were told. It seemed it was just an allergic reaction to something. Drew would break out from head to toe with red, raised, variegated-sized weals.

Randi and Brooke continued with their chronic illnesses and had developed numerous other, ever-increasing symptoms and problems that were unexplainable and didn't respond to treatments. It was obvious from infancy Randi, like Brooke, suffered from gastroesophageal reflux. Because Randi and Brooke were so close in age and both had so many problems, I decided to wean both of them from their bottles at the same time. Randi would wake every night and the only thing that would get her back to sleep was a bottle or two of milk. It seemed to soothe her. We didn't know if she was waking due to her daily, traumatizing doctor and hospital experiences, or if it was due to her gastrointestinal problems. Randi was two and a half years of age at the time.

Just a day after taking her bottle away, my curly blonde haired little girl came to me pulling her shirt up pointing to her abdomen and saying, "Mommy, my 'tomach hurts right here."

I still have a vivid picture of this ingrained in my memory. Her stomach hurt from that day on in the same place. By three and a half years of age, Randi was vomiting regularly after meals. It was quite a fiasco when we were out eating. We knew the signs to watch for. When they began, I would make a mad dash for the rest room, Randi in tow.

Brooke's extreme irritability followed her into her toddler years. She was a very challenging child, to say the least. We wondered if we would ever get through to her. It would take many repetitions in order for Brooke to learn each simple concept. I'll never forget my mom trying to teach Brooke how to eat a piece of fruit. She would watch over Brooke as she was eating, but no matter how many times Brooke had been reminded prior to and while eating not to eat the pit or seeds, when my mom turned her back for a second they would be ingested. Brooke would eat anything. While grocery shopping one day, Brooke lagged a few feet behind. When I turned around to see where she was, I caught a glimpse of her stashing something behind her back. I questioned her as to what she had behind her back. Trying to hide the fact that she had something in her mouth, Brooke replied with a side-to-side shake of her head. I knew better; she had green remnants sticking out of the corners of her mouth. After inspecting her hands, I realized she had grabbed a piece of lettuce from the veggie display. Another day, I picked Randi up from pre-school and placed her art project in the back seat. It consisted of painted popcorn glued on a piece of paper. The next minute, when I turned around to check on the kids, Brooke's mouth was every shade of the rainbow.

Drew, despite being the youngest, was usually the first of the three to do most things. He was the inspiration for his sisters many times throughout their lives. When the day came to remove Drew's training wheels from his bicycle, Randi protested and demanded to have hers removed as well. I was very hesitant and didn't think it was a good idea. Of course their helmets were strapped on. My kids wore helmets for every sport before it was the thing to do. Randi learned to ride a bike despite the fact that neither she nor Brooke were able to ride a tricycle because of their poor coordination, low muscle tone, and overall generalized weakness.

Granted, every one has to get out of Randi's path when she rides her bike because she is legally blind, but she can ride a bike without training wheels! Luckily, as the girls got older they realized they shouldn't always do what their brother did!

By the ages of five and six years, we finally decided to take the girls to Boston Children's Hospital. Our pediatrician and I had entertained the idea of having the girls evaluated at a major children's medical hospital for some time. Again, I had no desire to go to another city. This would be a four to five hour drive from home each way. I couldn't imagine how, by simply going someplace else, we were going to find any further answers. How could doctors only five hours away know any more than the multiple specialists we currently were seeing? The multiple specialists we had seen in Albany, our local big city, had already given up, saying we would probably never know the reason or cause for the girls' multiple symptoms and problems. We had made trips on a regular basis to these specialists. I can't emphasize enough to anyone who is reading this to go out of your local area to the biggest and best places immediately! If they don't have answers, as hard as it may be to do so, keep looking and trying!

I kick myself for not doing so earlier. It was mostly because I was afraid to go to a big city. I was so intimidated on our first several medical trips to Boston, I didn't even dare leave the hotel room other than to take the shuttle to the hospital.

I remember our pediatrician asking me, "Did you get a chance to see such-and-such while in Boston?"

I told her that I didn't dare leave the hotel. In fact, the first time we thought about going someplace in the city, we inquired at the hotel's front desk and were told to take the subway. The subway! I was not going to take the subway with three little kids, not after all the things I'd heard and seen on the news about subway stabbings, shootings and killings. I had never been on a subway. I asked the person behind the desk at the hotel, "Is it safe to use the subway, especially to bring kids on?" The people at the desk laughed at me. It was evident I had not gotten out much in my life! We did it, and other than the obvious inconvenience of traveling with three young children, it was fine.

We soon learned taxicab rides were far more hazardous to your health than the subway. Some of the drivers were downright crazy! Randi will never forget the time her seat belt would not unbuckle. The cab driver was angry and in such a hurry. He was yelling and screaming for someone to get scissors to cut her out of the belt. When no one produced a pair of scissors, he ran into the hotel's kitchen and returned with a large knife to cut the belt. On another occasion, while I was busy helping the girls out of the cab and safely onto the sidewalk, the cab driver took off with the door still open as Drew was getting out of the cab. That's not even addressing how they actually drove or the language they used!

Despite things being as bad as they were and the circumstances that brought us to Boston, I can honestly say I grew to love Boston. It became our home away from home for the seven years we made the medical trips, on average once every other month, for a week or more at a time. We made the best of a bad situation. The cities in which the major medical centers were located brought us opportunities. Certainly, my children's lives would not be sheltered as mine had been. They were exposed to a diverse amount of culture, places and people. Without doubt, this was a positive thing.

Some of my fondest memories are of watching my children attend the nightly shows that were performed at B.CH.H. After a long day of one doctor appointment, test and procedure after the other, we would often attend the shows. The kids looked forward to it. My children were the perfect age. They loved the shows and the interaction that was part of it. There were storytellers, singers, instrumentalists, or magicians. My three were not shy. They volunteered and interacted fully. It was proof to me that all the bad things in my children's lives were outnumbered by all the good. No matter how sick they were or how much pain they were in, this always brought a smile to their faces. There was happiness and the chance to forget health issues, if only for a short period.

Our first medical trip to Boston was a unique experience. My mom went with the children and me. We were lucky enough to find our hotel on the first attempt. After arriving at the hotel, I parked and then either took shuttles, taxis, subways or walked where we needed to go. No driving around in the cities any more than necessary. No matter where we've gone for medical trips, I've used the same technique. When we walked into the first hotel room, there was a huge full wall mirror on one of the walls.

Brooke said, "Hey, we have two beds, no wait three beds; well actually, there are four beds."

I will never forget that. All the rooms were the same, so every time we returned we'd joke about it. We stayed at this hotel, the Howard Johnson near Fenway Park, every single trip. The manager, shuttle bus driver, and other staff got to know us.

That first night, I was tempted to bring Randi to the emergency room at Boston Children's Hospital. It was almost unbearable to listen to her crying and her repeated complaints of throat and abdominal pains. I waited for morning knowing we were scheduled to meet with specialists who, hopefully, could finally help us. Brooke didn't have a much better night.

My mom said in the morning, "I don't know how Brooke is still alive. I woke up in the middle of the night and the only part of Brooke's body still on the bed were her feet."

Brooke was sleeping, hanging from the bed upside-down, with her head on the floor. Drew was found somewhere else on the floor. All of the kids had difficulty sleeping.

The first doctor we met was a birth defects specialist and coordinator of services. She was a pediatrician first and foremost.

On our first medical trip to Boston Children's Hospital, only Randi had appointments. We made mention of Brooke having similar problems but wanted to stay focused on Randi's multiple problems. We didn't want to complicate matters. Also, we weren't sure the girls' problems were related, although we felt they probably were. We also met with an immunologist and gastrointestinal doctor.

Soon after returning home, our pediatrician called and said the blood work performed at Boston Children's Hospital documented that Randi had an immune deficiency. She then called our local hospital to set up infusions of I.V.I.G.G. (intravenous immune gamma globulin), immediately. Randi had a low overall I.G.G. level along with subclass deficiencies. We were excited to have finally found some answers and hoped the infusions would make a big difference with regard to Randi's chronic infections.

During our next medical trip, both girls were seen. Brooke, like Randi, was diagnosed with immune deficiencies and she, too, began I.V. I.G.G. infusions. The girls also saw a geneticist even though they had previously been seen by one in Albany, N.Y. A simple blood test was done, and based upon this we found both girls had a one-of-a kind chromosomal anomaly. The first geneticist we saw could have done this same blood test! He pondered the blood test but made the decision not to put the girls

through the traumatization of a simple blood draw. Yes, it would

have been a little traumatic, but they had been suffering from

multiple problems and were already traumatized by them.

The first geneticist said, "They're too cute to have a

chromosomal anomaly."

They were all cute blonde-haired, blue-eyed kids described

by our pediatrician as three peas in a pod. Their looks should not

have been a basis upon which to make such a decision, however.

Randi had been traumatized at a young age by all of the

tests and procedures with the prodding and poking. I am sure her

lack of vision only added to her fears and inability to cope. (When

children are very young it's difficult for them to understand

reasoning. They just know they don't want any part of it all.)

On one occasion, during a visit to a doctor in Albany, Randi

bolted out the door and ran down the hall when the doctor entered

the room. I stood there stunned holding Drew who was just a baby.

I couldn't believe Randi had done this. She was only three years

old, at the most, and was legally blind. I thought she'd stop, but

she didn't, so I handed Drew to the nearest adult and took off after

Randi. It was a long, crowded hall. I hoped someone would stop her, but they just watched her running down the hall crying and screaming. My heart racing, I watched in fear as the door to the parking lot opened and closed again and again. It was just luck; the doors closed just as Randi reached them.

A chromosomal anomaly could, and we were told, probably did, explain the girls' multiple unexplainable problems. Further testing found the chromosomal anomaly was shared by the children's father, but Drew was spared. Most chromosomal anomalies are not compatible with life, the most common being Down Syndrome, where chromosome twenty-one is affected.

So many times people have asked me, "Why did you have so many kids if their father had this?"

We didn't know; no one knew! The children's father was delayed in school, but when he was younger, there was no testing or help available for learning disabilities. Neither was there testing for chromosomal anomalies. There were a number of family reasons as to why he had difficulty. Certainly, these could have been an explanation and therefore no further cause was pursued. He didn't have the other multiple health problems the girls shared.

We were told that the girls' father's chromosome analysis looked similar to theirs. When the girls' chromosomes divided further, because they were already abnormal, it caused a worsened presentation.

From what the technology available could determine, there was a partial deletion on the upper arm of chromosome nine, and what appeared to be an extra fragment of an unidentifiable chromosome attached at the deletion site. Several years later at the National Institutes of Health, with improved genetics testing, we documented that the extra chromosomal material was from the upper arm of chromosome ten. Things looked up for the first time since the children's births.

Soon after we received the diagnosis, however, Randi began developing new and worsened symptoms and problems, both G.I. and neurological. At age six Randi developed intolerance to physical activity and would become lethargic and pale. She would vomit, have increased throat and extremity pains after only several minutes of exertion, and would not recover fully for days after these episodes. This is when we decided we would have to either start using a stroller or wheelchair or stop doing things. Randi had just received a new bike for her sixth birthday, when what she really needed was a wheelchair. From that point on, Randi always had new-looking white sneakers due to her inability to engage in physical activity of any sort. She was able to go from place to place in our house, but anything else increased her symptoms and brought on the episodes.

At the age of eight Randi became unexplainably, totally urinary incontinent and had to wear diapers for approximately one year. There was something different or worsening every week it seemed, and we never knew what to expect next. Brooke followed Randi's path just as if she were her twin. Then, to everyone's shock, at approximately six to seven years of age, Drew started having health issues strikingly similar to his sisters. This was the same age at which Randi's overall health had taken a major nosedive. At first, I kept Drew's problems under wraps. After all, Drew didn't have the chromosomal anomaly his sisters did. I couldn't afford to raise doubts or complicate matters as the girls' problems continued to grow in number and severity. When I could no longer turn my head to his health concerns without allowing him to suffer, others including his teachers, began to notice and expressed their concerns. I was afraid the doctors would think I was crazy and would stop trying to find answers for the girls. As I had dreaded, that's exactly what happened!

Following an examination of Drew for his G.I. problems, our pediatrician asked Drew to sit in the waiting room and wait for me. I couldn't imagine why she would be asking him to do this. I thought perhaps she was going to give me bad news about our coordinator pediatrician at Boston Children's Hospital who had cancer. It was bad news all right, the worst imaginable, but it wasn't concerning our doctor.

Our pediatrician said repeatedly, "I don't want to do this; I wish I didn't have to do this."

Of course, I don't remember the exact dialogue word for word which ensued, but it went something like this: "I've thought this over and talked it over with other doctors. You can't take the kids back to Boston Children's Hospital. I received a phone call from one of the children's doctors at Boston Children's Hospital and one of the other doctors there is making accusations. If you go, we fear you'd be met at the doors with social workers and security and the kids would be taken from you."

All I was hearing was: This is it! Everyone is giving up on my children! I looked at our pediatrician and said, "They can't give up; not now! You know they are suffering terribly, and every day things continue to worsen for them." I couldn't help but break down. I was in despair. Now we had no help and no hope! We talked and talked. I was so frustrated! I remember making a fist, and in slow motion, hitting my leg again and again. Despite the fact that our pediatrician feared losing her medical license by maintaining an alliance with us, she assured me she would not give up on us. She also gave me a much-needed and appreciated hug. She then reiterated how sorry she was for having to do this to me. She said she knew what it was going to do to me, but she had no choice.

I was numb and felt as if I were almost in shock. This was very traumatic, akin to a death in the impact it had on me. It never even dawned on me until much later exactly what I was being accused of and how serious this could be. Not only was I suffering daily, my heart breaking, watching my children deteriorate and suffer, but my mothering skills and personal integrity were being questioned. I am very proud of my commitment to my children. I am an honest, down-to-earth person with values I take seriously. This doctor's action hurt me greatly in more ways than one. Before I left the pediatrician's office, I said, "We are still going on our upcoming planned medical trip." We had to!

Brooke was scheduled to have her tonsils removed at our next visit to Boston Children's hospital. She had been very ill for several months with severe, chronic, strep throat infections. She had been on I.V. antibiotics, daily shots of Rocephin, and nothing had helped. I was not going to let my children suffer unnecessarily. I could fight this and I would! I had nothing to hide or worry about! The only thing anyone could ever accuse me of was loving my children more than anything else in this world! My children's lives were at stake. The doctor who caused all this uproar obviously and truly didn't care about my children's health or well-being. I thought about it and decided the best thing to do was to have it out with whichever doctor was behind these allegations.

Too ill to work at the hospital, our coordinating pediatrician in Boston called me from her home. She explained how this situation came about, how it got to this point and which doctor was at the root of the turmoil. In her absence, without other doctors present who were familiar with the children's cases, their G.I. doctor brought about allegations at a group discussion. He voiced his theory as to why all three of my children had multiple G.I. and other problems he couldn't explain, and for which a gamut of medications didn't help. He suggested I was causing my children's problems and that I enjoyed bringing them to the hospital to have tests run on them. (Doctors at major medical centers hold gatherings periodically to discuss and brainstorm on tough or strange cases.)

We had also been contemplating Randi and Brooke having a fundoplycation for quite some time to alleviate the symptoms caused by their severe, unresponsive, erosive esophagitis. A fundoplycation wraps a portion of the upper end of the stomach around the end of the esophagus in an attempt to stop gastroesophageal refluxing. This was a serious, invasive, irreversible surgery with risks involved. Drew had begun seeing this doctor for the onset of new G.I. problems several months earlier. This could and should have been handled much differently, in my opinion! I know if I were a doctor and had suspicions of this nature, I would first pursue other avenues before causing trauma or possibly further endangering a child's well-being by my actions!

We had seen numerous gastroenterologists previously. They never had answers or could help us either and had given up, but they didn't do this to us. Doctors have the right to inflict this horrible nightmare on a parent and child on a whim. There are no consequences nor are they reprimanded in any way if their actions are wrongful. I can tell you that if anyone tried to take my children from me for whatever reason, I would have fought them to my death!

There were numerous individuals involved in our lives that were also dealing with the children on a daily basis and had been for several years. When I returned home with Drew from our pediatrician's office the day I was hit with the bad news, the girls' home health care nurse was there administering the girls I.G.G. infusions. I reiterated the story of what had just happened. She couldn't believe it. Without my asking, their nurse and others were ready and willing to help. My small army and I traveled to Boston Children's Hospital for a meeting with the G. I. doctor to discuss the situation: the girls' teacher for the visually impaired, the nurse from our home health care agency, and the girls' school psychologist. Before we went into the meeting, our pediatrician at B.CH.H.

guaranteed me her office staff had strict orders from her to watch my children and make sure no one touched them or took them anywhere. Both she and our geneticist were present. Our local pediatrician was on the phone line teleconferencing. The G.I. doctor didn't seem to care what others had to say; he maintained his opposition to my wanting to help my children.

Upon exiting the meeting, we all huddled together.

One of the troops immediate spontaneous response was, "What an arrogant little prick!"

Everyone agreed.

Even our Boston pediatrician had raised her voice, reprimanding the G.I. doctor during the meeting, telling him to stop and saying, "That's enough!" at least once.

During the meeting, I consented to a psychological evaluation by one of the psychologists on staff at Boston Children's Hospital. I didn't care; I'd jump through as many hoops as necessary to ensure my children continued to receive the care and help needed. It was somewhat embarrassing for all the doctors to know every aspect of my personal life, but I didn't care as long as it would put an end to this nonsense. From that day forward, I told myself I couldn't and wouldn't worry about what doctors or anybody else thought of me in my efforts to be an advocate for my children. (I am a somewhat shy and timid person by nature, but I will do what ever it takes to help my children!)

Brooke did have her tonsils removed, and the children followed-up with their other specialists, though it was evident there was still tension and controversy. It was clear that the majority didn't want to do any further testing or procedures on the children. Knowing this, I still had to push onward, continuing our quest for answers and help.

We returned a few months later in May of 1994 for a follow-up medical trip to Boston Children's Hospital. I knew the odds were against us with the new attitude towards the children's medical care. I wondered which doctors believed me and which actually still thought I could be at the root of my children's problems. Would this be a totally wasted trip? I felt it would be even harder to start all over with new specialists somewhere else. I didn't know what to do but decided to give it one more chance.

I knew we would not be home for Mother's Day due to this trip. We had to be at the hospital all week for appointments and then stay the weekend for one last appointment on Monday morning. I hated the thought of this and questioned whether it was even worth it all. Mother's Day is my favorite celebration because it celebrates my love for my children. Randi, Brooke, Drew, and I did some fun, special things together. As always, we tried to make the best of a bad situation. We went to a real Italian restaurant in the little Italy section of Boston for the first time. It was nice but all three had difficulty eating and became sick as usual. Each child picked out a card while we were in Boston and presented them to me on Mother's Day. Then we went to the hospital for Sunday mass. We had breakfast prior to this, which, like clockwork, made all of the kids sick once again. It was obvious to the priest watching Randi lay across my lap writhing in pain and discomfort, that she was suffering. The priest even looked at Randi when asking for prayers for the ill and suffering. Drew volunteered to read the gospel that morning. Although he was nine years old, he looked to be half that age since he was very small for his age. The priest wasn't initially convinced that Drew would be able to do this, I could tell. Drew made me very proud. He read as well as any adult. I was fighting back tears all through the mass. I had so much on my mind and was struggling with so many emotions. It was evident to me by the end of this week that the doctors at Children's Hospital were no longer willing to help my children. It was at that point in time, at mass on Mother's Day, that I made the decision never to return to Boston Children's Hospital. What we would do and where we would go next was equally as frustrating a thought. We'd have to begin all over again. This would be an enormous undertaking.

Only in retrospect can I now say this was my best Mother's Day ever! At the time, I thought it was the worst. I couldn't imagine how anything positive could have come from such a hopeless situation.

Things truly do happen for a reason, I've learned! I have also learned that worrying can never do any good; it can only make things worse. If I could just employ this strategy when I am in the heat of things, I'd be O.K. It's one of those things easier said than done. Surely I am not the first nor will I be the last to find these time-honored phrases exist for a reason.

CHAPTER 8

THE BAD, THE GOOD, AND HOW WE MADE IT

I'd like to not only dedicate this chapter as a means of sharing with you all of the good experiences that have come from living and dealing with mitochondrial disease, but also to vent many of the frustrations we have encountered. I do believe in the saying, "What doesn't kill you only makes you stronger." I never wanted to let all the bad things connected to struggling with this disease to have a negative affect on my children. I told myself, early on, I had to look for ways to balance the bad with the good.

In addition to my children's chronic pain and suffering, dealing with the doctors has been the worst part - for me anyhow. I am sure we don't hold the record for being seen by the most doctors, but we've seen hundreds. For sure, our situation and circumstances are relatively rare and unique in nature, which only further divides the good doctors from the bad ones. Addressing my family's health has put every doctor to the ultimate test and brought out their true colors. It is not my intent to discredit any doctor.

I can't, and I am not blaming them for being unable to help us or give us answers due to their inability to do so up until several years ago. What is maddening is doctors' continual lack of knowledge and understanding that currently exists! Doctors only know what they are taught in medical school and gain further knowledge through hands-on experience in their everyday practice of medicine. Ninety-nine percent of all the symptoms and problems our family has never fit into those nice, neat, black and white categories and realms of medicine, and now we know why. My children seemed to defy medical science, or at least initially it appeared that way to the doctors who had yet to hear of or deal with a mitochondrial disease. We are very fortunate to live in a time when mitochondrial diseases are finally diagnosable. I can't imagine going through what we have been through and never finding the answers, like generations before us.

Medical trips to see specialists usually brought about controversy concerning the severity and reality of my children's multiple problems. It was always a struggle just to try and convince our doctors to believe what the children were going through every single day was real. I was just a mom. What did I know? I was wrong and they were right just because they were the doctors. Parents know their children intimately, inside and out. We know when something is wrong. We know when our children's cries for help are real. I could never have helped my children or gotten to the point we're at now if I weren't one hundred percent sure I knew my children. One of the greatest inventions ever would be a device that could temporarily transfer the patient's symptoms to his or her doctor. This would enable the doctor to experience what the patient does! When I've shared this concept with doctors, they don't agree.

Our pediatrician would say, "Those other doctors aren't here in the trenches every day with us. That's why they don't understand."

The girls' immune deficiencies and dysfunction, as well as all of the kids' multiple gastrointestinal problems, were always points of contention and frustration.

Doctors have stated, "That's impossible!" more than once when describing symptoms, infections, reactions, diagnoses and test results, only to find out later that it was indeed possible.

After explaining to us what the normal case scenarios are for something, my children's pediatrician has learned to add, "But, for your kids, anything is possible."

There were times when nothing was being done, and the suffering was intolerable. At these times, I informed the doctors if they didn't do something, I was leaving my child with them. They would surely find the answers then quickly! Sometimes, I'd swear I was never going to have anything to do with a doctor ever again! I actually did have times when I couldn't and didn't deal with doctors for months at a time. I just didn't need, nor could I handle, the added frustration this brought. I have guaranteed our doctors that I would gladly go away and never bother them ever again if my children's problems went away.

I learned early on not to trust a doctor just because he or she was a trained medical professional. People often think of doctors as gods, and they are not. Most people trust them blindly and never question their decisions or actions. People are often intimidated by doctors; they are only human like all the rest of us. They all have their own very distinct personalities, morals, values, integrity and qualities that go into defining an individual. Doctors just happen to have chosen the field of medicine as a profession. I've come to these realizations having dealt with doctors both professionally and as a parent struggling to find answers and help.

Probably the one thing that I've found most upsetting and concerning when dealing with doctors is the number of times they accuse the patient, or in our case, the parent, of being depressed and/or crazy. Just because a doctor doesn't have an answer doesn't give him or her the right to accuse or lay blame for the problem on that individual! Granted, there are a few cases when this may actually be the case. I realize this, but it is done too frequently and without justification the majority of the time. I not only experienced it personally and repeatedly, but saw it every day from the professional side of medicine. A person isn't helped by misdiagnoses. It only causes them a host of further problems and prolongs their efforts to receive proper diagnosis and treatment. It's sickening! What about those individuals who can't and don't question the misdiagnoses? I've always respected and appreciated the doctors who told us they didn't know how to help us.

We've learned that the best doctors may not always have the answers but will openly admit this. They may not be able to cure the problem. However, they will stick with you through the thick and thin of it all and never give up. They will listen and trustingly work with you in an ongoing effort to try to find the answers and potential help, rather than placing blame for the problems. You don't always find the best doctors exclusively at major medical centers. We have had the pleasure of finding and working with a few of the best right in our local area. Having technology available makes a difference in even the best doctors' ability to care for you and help you. For this reason, having a good doctor at a major medical center is always a plus, especially when you're dealing with a rare medical condition. You are far more likely to be diagnosed if you suffer from a rare disorder at a major medical center because they see more of these cases. There are good and bad doctors everywhere. A lot of what goes into finding a good doctor or the right doctor is just luck. You may be fortunate enough to find a good doctor and then be referred by that doctor to others. This doesn't always work out either.

We were once referred to the best available local medical center specialist. Months later we learned from the nightly news he was a child molester! I didn't like this doctor from day one and felt he was not capable of handling my child's case.

After briefly examining Brooke, he stated, "I think she's depressed."

Initially, he refused to perform an E.E.G. on Brooke. A number of Brooke's other doctors and I strongly felt one was warranted due to her increased neurological symptoms. When he did finally agree to the E.E.G., the report generated conflicted with the actual testing performed. There was an entire paragraph detailing findings of a portion of the test that was never even done. I therefore questioned the results. I asked this doctor to inform our insurance company that my child needed to see another specialist, out of state, where our other specialists were located. We had fulfilled the obligation of first seeing the best available local specialist, per insurance protocol. This scenario led us to my children's current neurologist / mitochondrial specialist. As a result, we not only found Brooke had seizures, but also the answers we'd searched her entire lifetime for!

How about when doctors can't get along? We experienced this also. It has an impact on the patient's care. We sent information to one of our doctors, hundreds to thousands of pages of past medical histories, reports, and test results. Another doctor intercepted it and would not give it to the doctor for whom it was intended. I spent days searching through my files, making new copies, and spending money on shipping. What gave this doctor the right to withhold this information? I am sure all of you know, even when an issue or problem is black and white, doctors don't always agree. A person can go to several doctors for the same problem and receive a different opinion from all of them.

First and foremost, you need to educate yourself. You shouldn't rely solely on your doctor for your health. Your health and well-being is a combined effort. I was amazed when I worked in the emergency room at the number of people who didn't know the names of their medications, what the medications did, and why they were taking them. A person who has knowledge and understanding of good basic nutrition and health will not only be healthier, but will be better equipped to work with their doctor. No one can know everything, doctor or patient. I have enough knowledge to question some things sometimes and have found it was a good thing I did. It worries me to think about the times I don't know enough to ask questions. Are we missing things or doing something wrong? I probably would not be as educated regarding nutrition and health had I not been forced to deal with it to the degree I have.

Natural medicine was an avenue I chose to explore when none of our doctors could help us, and when all three of the children's conditions began to worsen once again post-Neocate. I have a great deal of respect for and favor natural medicine and its use of grass-roots, noninvasive, natural approaches to healing versus man-made chemical concoctions that often bring about more problems. I am therefore very happy the majority of the medications employed to treat mitochondrial diseases are natural supplements. They target the root of the problem, at the cellular level, naturally. Unfortunately, supplements are not covered by insurance companies, as they are not prescription drugs. Often, numerous combinations are needed which are very costly, especially when a whole family is affected. In people suffering with mitochondrial diseases, the supplements are needed every bit as much as prescription drugs. They are not merely supplements meant to enhance health; rather, they are necessary to ensure ordinary function and prevent actual diminution in quality of life. There are a number of supplements we have documented which provide some help and relief, but we can't afford them. There are more supplements we could and would like to try, but when we can't afford the ones we're now on, that's impossible. Natural medicine is becoming more and more popular these days, but there is not big money in it for the pharmaceutical companies so it doesn't get the attention and respect it deserves. It's a form of medicine which requires individuals to educate themselves and play a vital role in their health care. Nutrition is the foundation of everyone's health, whether your cells are normal or not.

Everyone's cells are affected by what we eat and what is in our environment. Our family purchases and consumes mainly organic foods in an attempt to ensure we are getting as much nutrition as possible and void of the things we don't want.

A patient or parent should follow up on their test results and reports. Ask for copies of them. This will also empower you and involve you in your health care. These reports and test results are all part of your permanent medical records and are used by other doctors as a basis for making other decisions involving your care and needs. To my surprise, I have found a great deal of inaccuracies in my children's reports. One doctor stated he removed Randi's adenoids when, in fact he told me, post the procedure, Randi had no adenoids to remove. Another recent report stated Randi had AIDS, versus the actual primary immune deficiency she suffers from.

There are times when tests and procedures are "messed up." One example was Brooke's recurring red, painful rash on her hands with exposure to sunlight. After the usual several months of no one knowing what it might be, I sent photos of the rash to our specialists. Our immunologist called to say he thought it was probably due to a chronic herpes virus. To confirm the diagnosis a biopsy of the skin including a vesicle, a fluid filled blister, was needed. Brooke and I knew this would be very challenging. Given a date and time for the biopsy, we had to plan it just right. Not only did we have to trigger the rash, but hope a vesicle would appear as well. It wasn't easy, and I am sure we received help from above. The odds of this ever coming together were next to nil, but somehow we pulled it off. Everyone involved in the process of obtaining the biopsy seemed to be very competent and careful.

They went out of their way to make sure the specimen was handled properly and arrived at the laboratory in the proper container. I was confident we'd soon have the answers. Everything had been executed perfectly. All we had to do was wait. Wait and wait and wait, the usual process that has always been involved following every test and procedure. This, of course, was in addition to the initial delay of scheduling an appointment and waiting months until the doctor could see us. The ball was dropped somewhere in the final step of the process, and the vesicles that were painstakingly removed without rupture were not cultured. Instead, only testing which used I.G.G., I.G.A. and I.G.M. antibodies solely for detection and diagnosis was employed. Brooke has no I.G.A, greatly decreased I.G.G. and I.G.M., and doesn't build titers to many things. All of the doctors involved were aware of Brooke's immune deficiencies and dysfunction. In fact, Brooke wouldn't have a chronic herpes viral rash if she didn't have immune deficiencies and dysfunction. How could this have happened? I promised Brooke that we would never again make her go through even bringing on the rash for testing purposes, let alone another biopsy.

That's why when I worked in the hospital and a patient was administered any test, I personally took care of the specimen from start to finish, especially when it involved one that the patient, wouldn't want to, or couldn't have repeated. There are several steps in the process when something can go wrong. First, all of the tests must be ordered correctly. Then the specimen must be collected and handled properly. There are certain containers and preparatory solutions for specific tests. The specimens must then be labeled correctly and received by the laboratory within a specified time frame. Then it's up to the laboratory staff who, in turn, have several steps of their own that must be executed accurately in order for the whole process to be considered a success.

I hate being reliant on doctors. I can't tell you the number of times I've waited days on end for a phone call from a doctor. I am glued to the house, prevented from accomplishing anything. Walking to the mailbox, vacuuming, or even making a trip to the bathroom guarantees this is when the awaited phone call will come. The missed call begins the long stressful ordeal all over again.

Often a doctor promises they'll call on a certain day and then they don't.

I don't mind advocating for and coordinating the majority of my children's medical care and needs, but sometimes I feel like the doctors rely on me too much. For example, I must regularly place several phone calls in order to address and resolve just one issue. I feel if I don't follow-up on everything (test results, appointments, procedures) nothing would ever get done or be resolved. I feel like we would never hear from our doctors again. It scares me especially with my decline in mental as well as physical abilities. My children may now be young adults, but they alone are not capable of, and in some cases not willing to do, all that it takes to advocate for themselves. It's the last thing individuals of this age think about or want to do. I have to say dealing with doctors, and all of the above-mentioned trials and tribulations associated with them, is in itself enough to drive anyone crazy! If you weren't crazy to begin with, it's very likely you will be by the time they get done with you! I, along with anyone else who's had similar ongoing experiences, have every right to be unquestionably, certifiably crazy!

Dealing with the doctor's office staff can be yet another frustrating obstacle. You can't or shouldn't judge a doctor solely by his or her staff. The down-right incompetence and lack of skills some of the office personnel repeatedly display is laughable. I know people have a bad day from time to time, and everyone does make mistakes, but when it's the rule rather than the exception there's something wrong. Everyone reading this that has had similar experiences and is laughing right now knows exactly what I am talking about.

As I complete the writing of this book, our pediatrician has informed us she is retiring soon. To date, numerous local doctors contacted have refused to take on our cases.

She has even admitted to us, "If you were to walk into my office today, and I didn't know you, and hadn't been through all of what we've been through since the children's birth, I would think you were crazy, and I wouldn't take on your cases either."

Our specialists are the only ones besides my children's pediatrician who understand and can help us. However, they are several hours away and don't have the time needed to deal with all of the day-to-day problems that require a doctor's input. They usually only see patients for follow up discussion of their disease progression and treatment options. I have resigned to making the five and a half hour, one-way trip without any second thoughts. I would travel anywhere to ensure my children receive appropriate medical care. Waiting months to see a doctor for acute problems is a terribly frustrating situation which further hinders care and treatments. We have doctors on both ends of the spectrum. Our specialists who deal with mitochondrial diseases laugh when we tell them of local doctors who have never even heard of mitochondrial diseases, and if they have, they know little or nothing about them.

We first experienced this phenomenon at B.CH.H. and have continually seen it at all of the major medical centers we've visited. The technology and knowledge that exists today at major medical centers won't drizzle down to our local level for approximately ten years or more. In most people, even when faced with a more common ailment, this cutting edge knowledge and the advanced treatments can make a tremendous difference in the options available and outcome.

To complicate matters, the children are now young adults and have aged out of their pediatric hospitals. Although we've been able, up until now, to continue seeing our specialists located in the pediatric hospitals, it presents a problem. My children can no longer have testing, procedures, or be admitted if the need arises. A few exceptions have been made, but we were told they couldn't continue. We currently have no doctors at adult hospitals that can provide these things. It's just one more traumatizing aspect to overcome. I feel like we've once again been abandoned, this time, by the only doctors that can help us and understand. It's taken the children's whole lifetimes to find and establish the relationships we now have, and we have to just give them up for this reason alone. Our specialists acknowledge we are not the only ones affected by this practice, and it is being evaluated and questioned. In certain places, they are currently setting precedence by developing a protocol to address this important issue. Meanwhile, the majority of the girls' care is on hold. As Murphy's Law would have it, their health problems haven't been this bad, for all three, since before Neocate.

That's the bad. Now here's some of the good. As long as there is some good, no matter how little, you can find the faith and the strength to go on. I've learned change isn't always a bad thing and if things don't work out it's for a reason, even if I can't see it at the time. God, and only God, knows what our future holds. I am sure what is to come won't cease to amaze me any less than the past has. I find encouragement and hope in the future regarding stem cell treatment. After all, despite the enormous amount of research and technology which goes into making it possible, stem cells are natural; they aren't foreign or man-made. Faulty components are being replaced with healthy components free of the previous defects.

Our pediatrician is the only doctor who has survived this rocky road with us since the children's births. She will always be our number one doctor, in my book anyhow! There aren't words to express all that she and other special individuals have done for us.

When the children were babies, she would say, "The children's angels forgot to leave their instruction manuals with us."

Our pediatrician always let me play an important role in my children's medical care. I had to have an outlet for all of my frustrations. I had to do something constructive with it. She encouraged me to constantly research and share ideas and suggestions. We have always brainstormed together in an effort to come up with adaptations and modifications to meet all of their unique needs. Doctor appointments for my children are not typical; rather they are strategizing discussion sessions, as much as anything else. I've extensively documented the children's multitude of symptoms and problems by keeping a daily journal for each of them. This allows me to compile medical histories and updates to share with our pediatrician and specialists. I've also helped in coordinating their care, ensuring all doctors have copies of all lab results, reports, and that they are accurate. There would have been no way any doctor alone could have done all it takes to do what we have done. A parent must get involved when faced with a situation like ours. There has rarely been a day over the past twenty-three years when the children's pediatrician hasn't had to do something for us. If she weren't examining one of them in her office, then she was ordering some test, reviewing some report, consulting with one of our specialists, asking some insurance company for coverage, filling out forms, or giving me a call to discuss things. I am sure there have been times throughout the years when she has asked herself why we chose her. In the beginning, we randomly chose the group she was part of. Nobody could ever have guessed what was to be. As time went on, I decided we needed to choose one of the doctors in the group to see exclusively. With so many chronic problems, I felt we needed continuity. Our pediatrician was there the day of Drew's discharge from the hospital after his birth. I caught a glimpse of her casting a genuine, cherished, reminiscent look towards us. That day I made my decision as to the doctor I wanted to care for my children; this memory was what swayed me. I have trusted her all these years with my most precious gifts. She hasn't had all the answers and has been wrong sometimes, but I've been wrong too. She's had to listen to me rant and rave about doctors. The bottom line is she's always been there and has never given up on us! Sometimes, just talking to her has made me feel a lot better and helped me go on one more day.

Specialists would ask us if our pediatrician was still involved in the children's care and then state, "She's a really amazing person to have stuck with you guys all this time, throughout everything."

Our pediatrician has been a constant, solid source of support and help we could always count on. I know how very truly blessed we have been to have her.

I would like to make mention of a few more doctors who will always have a special place in my heart for their contributions, compassion and support. I had wanted to mention a specific handful of doctors by name in an effort to thank them. However, for a number of reasons, I decided not to use any of our doctors' actual names.

The first doctor we met at Boston Children's Hospital, the coordinating doctor of my children's care, was another very compassionate, motherly woman who went out of her way to help us on more than one occasion. For instance, we had gone to B.CH.H. for a weeklong medical trip. Brooke's tonsillectomy was scheduled for the following Monday. Our original plan was to go home Friday after our last doctor appointment, and Brooke and I would return Sunday for her surgery. Well, Mother Nature had other plans. A major snowstorm hit while we were at the hospital Friday morning. It was predicted to be very big and it was. Our coordinator doctor called me from her cell phone in her car. Like everyone else, she was trying to get home before the storm got any worse. She told me it was too dangerous for us to try to make it home. We discussed the possibility of one of the young college girls who was working at the hospital in her department to care for Randi and Drew while I stayed with Brooke in the hospital. It just so happened this young lady had also worked as a counselor at the special needs camp that both Randi and Brooke attended. I was hesitant because of Randi's extensive medical problems and the fact that I didn't know this person. However, I had no other options available, and I needed to act quickly to see if this was even a possibility. I trusted this doctor, and she felt it was a good solution. She gave me the girl's name and phone number. I called her with my proposal. I would pay her a set amount of money by check to stay with Randi and Drew at the hotel overnight, and then bring them to me the next day at the hospital. The hotel staff was so accustomed to seeing us over the several years we had stayed there, every other month or so, that the girl was questioned as to where I was and why she had my children. The worst part of her job, though, I was told, was my son Drew. She told me Randi was no problem but Drew was a devil in disguise. He gave her a run for her money, annoying her in any way he could think of the entire time.

Unfortunately, this doctor developed cancer that eventually cut her life short. She had two young children. I remember her calling me from her bed at home one day to discuss my children's care. Her young son was trying to get her attention the entire time we were on the phone. She was one of the few doctors that believed in us and was supporting our efforts right until the end. I wanted to surprise her by visiting her at the hospital with the children following their Neocate miracle, but it was too late. She had already passed on.

Then there's the geneticist at Boston Children's Hospital who tested our family for and diagnosed the girls with their one-of-a kind chromosomal anomaly. I think we always feel greater attachments to those doctors who find answers. He became co- coordinator of the children's care at Boston Children's Hospital when our coordinating physician was taking medical leave. Based upon this genetic documentation, the girls were finally able to receive crucial services and financial assistance often denied until such proof was obtained. Our geneticist left B.CH.H. around the same time the other doctors there had given up on us. We told him that we wanted to continue to follow-up with him, as he was one of the few specialists left who understood and could help us. It was one of my reasons for choosing Johns Hopkins. He relocated to National Institutes of Health in Bethesda, Maryland. I figured we could see him on our medical trips to Johns Hopkins, and we did for a few years until he moved on to California. I still keep him updated by letter at least once a year. When you find a good doctor you don't want to give them up. They are so few and far between!

Our gastroenterologist, as everyone knows from the first chapter, will never be forgotten. He may still not be able to tell us why Neocate has had such an impact on my children's lives, but he was the man responsible for placing my children on it and helped us obtain it. He is a very humble man, and to this day states he never did that much for us. For many years post Neocate, our visits with him were very uneventful because Neocate had completely resolved all three of the children's multiple gastrointestinal issues. Follow up is important even when everything seems to be going well because when, or if, you need the doctor's help they are up-to-date and familiar with your case.

During one of our last conversations with him, in preparation for transitioning to adult doctors, he told us, "I can never fully turn you guys loose. You're a special family and in dealing with you, one can never say anything is for sure."

A million thanks go out to all the nurses and therapists throughout the years who have not only delivered compassionate, competent care to my children, but have also given us support in so many ways. We developed a special bond with these individuals seen on a regular basis, because they got to know the children, their complex problems, needs, and the changes which constantly took place that impacted their lives.

Medicaid and S.S.I. are supposed to help you, but sometimes only add to your list of problems and frustrations. You face trying to acquire help for something that no one has ever heard of. The medications, tests, and care needed have not been needed or asked for before. I must give Medicaid a great deal of credit and thanks, however. Medicaid has been there for us numerous times. We would not have been able to meet the children's medical needs, financially, without them. I feel part of our latest battles not only stem from lack of knowledge of this disorder but cutbacks in funding.

A problem associated with having Medicaid is the stereotyping and stigma that accompanies being on Medicaid. I can tell you for a fact; its real and it happens! I've dealt with it from both sides of the coin. In my professional realm, nurses and doctors would often decide a patient's needs and the course of care based on whether they had Medicaid or private insurance. I would say to them, "You know my children are on Medicaid; not everyone on Medicaid is a low-life. You'd be surprised why some people are on Medicaid."

They'd reply, "That's different."

When you pick up a person's chart and automatically stereotype them before ever meeting them, how do you know what the individual's circumstances are? I can tell you from a patient's perspective, you are treated differently based upon whether or not you have Medicaid exclusively. I actually like to say, for the most part, you're not treated at all. No one wants to accept Medicaid and generally will offer you only the most basic of care, if any at all.

The U.S. health care system currently leaves a lot to be desired. Health care is a most basic and essential need for everyone. It shouldn't only be available to the lucky few who have access to a policy or can afford a policy. This subject, although it's extremely important to our family, is another issue, another book.

As for S.S.I., we are also grateful it exists. Our family found its rules and regulations further strained our efforts to survive financially. Once the girls had been diagnosed with a chromosomal anomaly, they were entitled to S.S.I. benefits. However, when the children's father or I had the opportunity to make a little more money, which was rare, we had to pass it up or the girls would lose Medicaid. Though Medicaid was usually only a secondary source of coverage, it was priceless. The girls' expenses after primary insurance were still hundreds of thousands of dollars per month. Medicaid also paid for our transportation costs for out of state care. There were times, between coverage by health insurance policies, when the girls had exhausted a policy and Medicaid was the primary insurer. We had to stay poor to get the needed help. (This is one of those times if common sense was employed as part of the equation, like so many of the issues in today's world, there would be a lot less chaos and people would be a lot better off.) You would have to be a multi-millionaire in order to be able to deal with this on your own. I had a doctor ask me how we made ends meet. He had a special needs child and found it difficult financially. Now that the girls are over twenty-one, they are discouraged because if they are able to occasionally earn small amounts of money, all but a few dollars is deducted from their S.S.I. The number of hassles involved in notifying the agency and showing proof of the earnings makes it pointless. I also realize, but will not let it stop or deter me, that my writing this book will likely result in our becoming worse off financially than we are now.

At this time, nearing the completion of my book, we are totally reliant on and a financial burden to others for our needs. It's terribly hard to come to grips with this, on top of everything else, when left with no other choices or options. Believe me when I tell you I can't imagine anyone wanting to go through this, or putting their family through it, unless they absolutely have to.

I recall someone at work, a young person, saying, "Oh, you're so lucky you don't have to work anymore."

I wish I could work! I have experienced the full range from normalcy to being non-functional due to mitochondrial disease. The difference between the two is like night and day. If the doctors can ever come up with a cure, my children's lives would be so different. You can't explain normalcy to someone who has never known it. Even though my genuine aspiration is to help others by writing this book, I am also hopeful I may once again gain control of my family's finances, and in turn, their well-being.

The school was another area where, for at least the first several years, daily battles were waged. Randi had obvious, outward dysmorphic facial features. This made it easier for her to be labeled and helped at the pre-school level. Not that a person's obvious deformities should ever be the sole basis for determining a person's needs or functional abilities, but it happens all the time.

Brooke, on the other hand we were told, was "too cute" to label. Therefore, she went without pre-school special education when, in fact, she needed it more than her sister did. We knew Brooke's developmental delays were worse than Randi's, but we were on our own to provide her the much needed help. When kindergarten rolled around, our pediatrician wrote a note to support special education for Brooke. Again, I was told by the school that Brooke was "too cute."

I was told, "Give her a chance and she'll prove you wrong. Let her go to regular kindergarten."

Well, I fought to get Brooke into special education, and by the end of that same school year, the school had drastically changed its tune. The same school personnel I fought with to get special help for Brooke were now stating that Brooke was far too severely learning disabled to attend regular school. After one year of both the girls attending the school's rendition of special education classes, I decided the girls would be better off in a regular school setting in their hometown school.

I attended an advocacy seminar offered on mainstreaming special needs children. There I found valuable information and individuals who could assist me in my efforts. Another mom who had already pioneered the system with her son and had experienced success was one of the key speakers. She not only gave me advice when I asked for it, but she actually came to a committee on special education meeting to inform our school district of the laws and of my daughters' rights.

She also recommended a neuropsychologist. He goes down in my book as one of our best doctors! Of course, the school always gave us "flack" when it came to our having our private psychological evaluations performed at the school's expense. This was just one of the many ongoing battles we waged with the school and won throughout my children's school years. This doctor was not only instrumental in identifying for the school what Randi and Brooke's neurological impairments were, but how they impacted their abilities to learn. A detailed report containing information the school needed to implement in order to meet all of Randi's and Brooke's academic and medical needs was compiled. The number one recommendation that made it all work and come together was a personal teaching assistant. The teaching assistant was to be utilized exclusively for the girls and their multiple needs. This was always a controversial issue early on, but became less of an issue once the school realized how great the girls' needs were.

We had to involve lawyers at least twice in our disagreements with the school district. On one of the occasions, the school district didn't want to allow me to audio tape the committee on special education meetings. I asked to do so in an attempt to put an end to the school's conflicting recollections of important issues that had been discussed at previous C.S.E. meetings. As time went on, both the girls' teacher for the visually impaired and their personal teaching assistant naturally became advocates for the girls. It no longer was exclusively my word against the school's. I always made sure these individuals were present for every C.S.E. meeting. Having other allies is always a plus in such situations. Having continuity with the same allies is even more powerful. The girls' teacher of the visually impaired first met and began working with the girls at the ages of four and five years of age. She worked directly with the girls on a twice-weekly basis at school and /or at home. Due to their extensive health concerns and needs, the girls were schooled at home as much as, if not more than, the time they spent in school. Initially the girls shared the same teaching assistant who began working with my daughters in pre-first grade. From first grade on, all three of the children were in the same grade.

A teacher was hired to tutor the girls and officially joined our team in fifth grade. I didn't pay much attention to her at first. That was a crazy school year. The kids had missed the first few days of school and then had to return to the hospital yet again the following week. There were several more medical trips that year as well. Luckily Drew didn't have any difficulties with learning and was able to keep up on his own. The new teacher became one of the three unforgettable, indispensable individuals that would be there for us all throughout the girls' school years and beyond.

They all became the girls' second moms. Each of them grew to know Randi and Brooke as intimately as they knew their own children. They were able to care for Randi and Brooke in both the school and home school settings. They learned Randi relied on those around her, especially at school, to read her facial mannerisms versus expressing herself verbally. It usually takes Randi quite a while to come up with her responses when she is asked a question. As she got older, Randi revealed to us it takes her time to process information in her brain. Randi always functions in a very slow, even-paced mode. She rarely demonstrates deviation in her expressions even when she is excited. They also learned Brooke will say or nod yes even when she doesn't mean yes, usually accompanied with a big smile. The two girls together were an accident waiting to happen. Simply walking down an empty hallway could result in an entanglement and injuries.

In spite of this, I could completely relax whenever the girls were in their care. I developed a trust, faith, and special relationship with each of them. Through a cooperative effort between the school district, the girls' classroom teacher, and teaching assistants, a program evolved in which home instruction was provided by school personnel. This mirrored the classroom agenda and curriculum so that when the girls returned to school they would be in sync with their peers. In addition to home schooling for several years, the school provided a separate classroom for the girls from November through May, the worst of the cold and flu season. Even then, when and if the girls attended, it was usually for only two to four hours per day. The girls weren't instructed in every subject, just the core ones like math, reading and writing, and a little bit of art, science and history thrown in from time to time. With the one-on-one teacher to student ratio, they each received intensive, focused, individually tailored nurturing and academics. I attribute this as one of the main reasons for the girls' academic successes. They both have accomplished more than anyone ever initially thought possible. It's the by-product of the continuity of those few dedicated, caring individuals involved.

An example of this dedication was when the girls' teacher for the visually impaired came up with the bright idea to enter a contest to win tickets to a Nickelodeon show. I informed her she had to promise to accompany us to the show if the kids won the tickets. The essay had to include who you would like to see slimed and why. "Slimed" referred to a green gooey substance popular and synonymous with Nickelodeon, a children's television show.

The girls' immediate response was, "Our doctors!"

With the help of this teacher, they compiled the essay and sent it in. They won! Off to Albany we went with a van packed with nine. The V.I. teacher chose bathroom duty. This consisted of ushering several kids to bathrooms located approximately a mile away from where we were sitting up and down several hundred stairs in the arena. She made this journey a few times during the show while I remained with the rest of the children. I reminded her that this was her idea. By the time the show ended, it was late. We faced a long trip home, so we pulled up to a drive-through and ordered the kids' dinner. It was dark and I was driving. I needed to be able to check on the kids so I yelled out, "Choke check time! Everyone check on your neighbor sitting next to you." The V.I. teacher started laughing and to this day we joke about that phrase. This was prior to Neocate use and my children had difficulty with their food getting stuck in their throats. I was worried one would choke and I might not know it. I am sure most families don't conduct choke checks.

Every year the girls were assigned a classroom teacher, like all the other kids. Initially, these teachers were very hesitant and intimidated when we met to discuss all of the girls' health problems, needs, and academic difficulties. However, each and every one of them ended up excelling in their efforts to help and support us! After getting to know Randi and Brooke, the classroom teachers all commented that they wished all of the children in the classroom were as courteous, caring and well-behaved as the girls. I truly believe and have always tried to ingrain in my children that no matter what your abilities or disabilities, the most important thing in life is to be a good person. I told them, "This will get you further in life than anything else." I've always said and truly mean it when I say I love Randi and Brooke just the way they are. They are very unique, special individuals of whom I am so very proud! The only thing I would change, if I had the chance, would be to take away their daily pain and suffering.

Transdisciplinary meetings were held at the school every two weeks in an effort to bring everyone together to discuss progress and any problems that needed to be addressed. The classroom teacher, tutor, personal teaching assistant, teacher of the visually impaired, special education teacher, a C.S.E. chairperson, and I all attended these meetings. The girls' individualized school programs were top notch and others tried to fashion their programs after them.

Randi, Brooke and Drew all graduated in June 2002 from Warrensburg Junior / Senior High School. The girls' teacher of the visually impaired and their tutor continue to be irreplaceable fixtures in Randi's and Brooke's lives post high school. These two dedicated individuals continue to support and advocate for our whole family and our needs. We all continue to lend support and help to Randi and Brooke as they explore their young adult world. Randi has a passion for writing and enjoys creating artwork to accompany her children's stories. When her health permits, she likes to share her work with young children in our community. She relates well with children. Composing poetry and listening to music are means Randi uses to cope with her constant pains and other limitations. Randi has great difficulty with handwriting due to fatigue, weakness, and poor coordination of her hands. Therefore, the girls' computer training has been a great aid in assisting Randi with her endeavors. It also allows Randi to explore the world and still have friends. Randi is homebound the majority of the time because of her chronic long-term infections and severe allergies. Multiple other symptoms and problems greatly limit Randi's abilities and options as well. She is very happy, however, despite it all.

Brooke is able to go out into the community a little more. She enjoys dance, music, and theater. Brooke actually is in two very different drama classes. She attends a touring theatre class at our local community college and takes part in a theater group of developmentally disabled peers. Brooke also enjoys working with young children. We are fortunate to have a very understanding, caring local pre-school teacher. She was Brooke and Drew's pre-school teacher and now allows Brooke to be her volunteer helper. Because Brooke is so normal looking, and her social skills are her strengths, people have no idea just how developmentally affected she really is. Many don't understand. Initially, they don't believe us when we tell them how limited Brooke is until they get to know her. It's definitely a double-edged sword in regard to her safety. Brooke repeatedly experiences confusion episodes which have a negative impact on her ability to function. Self-medicating, crossing roads, performing mathematical skills and composing written language all are difficult for Brooke. She also has severe visual perceptual difficulties. Although her visual acuity is only slightly affected due to her congenital optic nerve defects, Brooke's perception is greatly affected. She has profound difficulty with writing but can talk your ear off. I have found notebook after notebook with page after page in which she has tried repeatedly to get a few words or sentences written before starting it all over again. With each new attempt, she may lose or gain a word or two in the process. There will be a dozen pages or more of the same words and sentences. The computer has aided her some, but without another person's guidance, help and refocusing, Brooke can spend several frustrating hours and never get beyond a few sentences. The confusion episodes are most prevalent and are the worst for

Brooke. Randi and I experience them as well, but not as much as Brooke does. Brooke's are repetitive throughout any given day, and have been present for as long as she can remember. Brooke states, and I can attest that the confusion episodes seem to be triggered or made worse by trying to focus and or concentrate on something. The brain just shuts down. It's too overwhelming. The harder she tries to complete the task at hand, the worse it gets until she has to just give it up. That's easier said than done, depending on what it is she's trying to accomplish when the confusion episode hits. It's dangerous if driving or trying to cross a road. I can see why Brooke has had such a hard time with learning and completing tasks. If I had been affected from the beginning to the degree Brooke is, I wouldn't be functioning any better than she is. For whatever reason and a perfect example of the variability of a mitochondrial diseases, Randi's confusion episodes after starting Neocate are only experienced during menstral cycles. Somehow Neocate blocks or compensates for part of the metabolic aspects of the faulty mitochondria. During menses, Randi is also very-run down physically and is mentally drained. It's a stress on her body that pushes her beyond her tolerable limits.

Drew never had learning problems, but this mitochondrial disorder affects him in a myriad of other ways that ultimately have a negative impact on his ability to attend school and to reach his full potential in academics and in his personal life. Drew started out in life as a very confident, capable young man. Many of those who met him for the first time actually referred to him as a little man. I can still picture Drew on one occasion mingling and walking amongst our doctors as they conducted testing and were observing his sister. I called him aside away from the doctors but they said it was fine; it was so cute. I was proud of his lack of inhibition and the manner in which he handled himself. He has always been very articulate in expressing himself. This disease's cruelty has limited and robbed him of many of not only his own aspirations but those I had for him. Drew was always thought of and referred to as my "normal" child. Even when he initially developed symptoms and problems similar to his sisters', no one really understood because he was not affected developmentally. Many individuals with mitochondrial disorders "look" normal. Drew was blessed with a natural talent for music. He taught himself to play the piano at the age of ten. He has the ability to play numerous instruments, although trumpet and piano are his favorites. He has always found comfort in his musical ability.

Life is a lot of things. We go to school to prepare for life, but there is no course "Dealing with all this Stuff 101". We are never prepared for the kinds of things our family has been through. In these instances, although we sometimes wish we could gaze into a crystal ball for the answers, overall, it's best we can't. If someone had told me the story of my life, I don't think I'd have believed them. If I did, I'd probably be too scared to live it. By no means are we the first, nor will we be the last, to ever deal with unbelievable, extenuating circumstances. Many have experienced even worse.

The one phrase I hear most often is, "How do you do it?" It's true, the frustration can be indescribable at times.

One of my most soothing ways of dealing with chronic overwhelming stresses was building rock walls. In contrast to my life, which I have little or no control over, every rock in the walls was painstakingly perfectly placed and permanently set in concrete. I could work out my frustrations physically while doing something I loved. It just so happens our entire yard is made of cobblestones. You can't dig a shovel full of earth without coming up with a variety of rock. Ironically, because of health issues, I had to give up on this project and dream along with so many others. Looking at my unfinished landscaping project is one more constant reminder of the disorder that has changed my family's lives in so many ways. A friend recently reminded me that my soul is still very much alive and well. Although this disease may have dibs on my body and brain, it can never gain control over my soul!

Anyone who has dealt with extenuating circumstances knows there usually is no one thing or one person that makes all the difference. It's a combination of a few, very special constants along with help and support of many others sprinkled throughout your journey. Despite all of this, there are still times when you question, "How can I go on?" I want to share with you some more of these special people we've met and places along the way that have made it possible for us to go on.

One is Kevin Luibrand, the lawyer who stepped up to the plate and went to bat for us when no one else would or could. We will always be indebted to Kevin. Neither the children nor I will ever forget him or what he has done for us!

My children began dance lessons at an early age. Randi had to discontinue this activity because of her poor health. Brooke continued with dance for several years despite the challenges she faced mentally and physically. The instructor, recognizing Brooke's passion for dance, was patient and understanding. She gave Brooke private lessons when Brooke could not keep up with her peers. Initially, it seemed like an impossible dream since Brooke showed no potential in this endeavor. As Brooke continues to dance, she has become more skillful. Her early foundation in dance was incorporated into other favorite activities such as cheerleading and acting.

The Double H Ranch is truly one of the greatest places on earth and is located right in our own backyard, approximately twenty minutes away in Lake Luzerne N.Y. Our pediatrician has been the medical director there since its beginning in 1993. This alone should give you an idea of what a special place it truly is. It was founded by a local philanthropist, the late Mr. Charles R. Wood, and actor Paul Newman. Double H Ranch is one of two original special camps for children suffering from neuro-muscular or blood disorders that are terminal or chronic. The children who attend these camps cannot attend regular camps because of their medical problems. Randi and Brooke were only six and seven years of age the first year camp started. I had my doubts that their multiple medical problems could or would be addressed there. Only because our pediatrician was the medical director did I agree to let the girls participate. Our whole family has been touched by the spirit which resonates from this very special place. Around Christmas time each year, there is always a big party attended by campers, counselors, nurses, doctors and the many volunteers who make it all possible. Camp Double H also offers a winter adaptive ski program which Randi and Brooke have enjoyed. This will be Randi and Brooke's last year at camp. They aged out of the regular camper sessions but have been able to attend an alumni session for the past few years. I never could have envisioned how much Randi and Brooke would benefit, and what a significant, positive impact camp would be for them.

Another one of the girls' favorite activities growing up was the 4-H therapeutic horseback riding sessions. The girls loved the horses. Not only were they receiving beneficial physical therapy but building self-esteem in the process.

Numerous volunteers along the way have given of themselves so my children and many others could enjoy opportunities which would otherwise be impossible.

Coming from a very solid, loving, close family, I assumed all families were like ours. Once in the real world, I was shocked! My two brothers, sister, and I had the best mom and dad anyone could ever ask for! With this solid, strong foundation came high self-esteem. I thrived on competition. If someone said it couldn't be done, I'd do it just to prove them wrong. Never afraid to go after things I wanted, I felt I could do anything if I tried hard enough. Being stubborn, I would do what it took to get my way. With a love for sports, I was strong and tough enough to play them with the neighborhood boys. Besides playing sports, I enjoyed building things. Although these were then considered boy activities, I didn't care what other people thought.

I am glad God gave me all of these skills and strengths. They came in handy to give me the confidence I needed to face the obstacles in my life.

My mom and dad live right next door. They have also always, without a doubt, been a constant, unwavering source of strength, help, and support. They watch the kids and care for them financially without ever a question. They have consistently been the ones to bear the burden of all of our family's unmet financial obligations. My mom and dad have gone without so we could have things. Their own inability and circumstances has never stood in the way. They continually did whatever was necessary to help us out. They are middle-class Americans trying to make it like everyone else. My dad has always automatically factored in our needs with theirs. I can never repay them, short of winning the lottery or something else of that magnitude. They are the ones who have paid for my children's multiple, vital non-covered medications. Over the past several years, the cost each month has been hundreds of dollars. Even when they've given us all of what we need, they give us more. It makes me sick to take from them, but there are no other options. I must burden my mom and dad to obtain medications to help my children. We wouldn't have half of what we do if it weren't for my mom and dad. The added expense of maintaining a safe, reliable vehicle for our many long out-of-state medical trips is essential. My dad is a mechanic and has saved us a lot on vehicle maintenance and repairs. With the number of miles we drive, a vehicle wears out before we have it paid for. Then we have to start over again with more payments. I can't take the chance of missing an appointment we've waited several months for or a special procedure that can only be done at a specified time. I don't want to break down in a big city, be involved in an accident, or be put in a dangerous situation with an unreliable vehicle. We've come close a few times. Most of the time it's just the girls and I and the girls aren't able to do much. Transportation is just one more thing I don't want to have to worry about or deal with on top of everything else and don't have to because of my parents' help.

My family has been intimately involved in my children's lives since their births. We gather as a family to celebrate every birthday, holiday, and any other special event. Both of my grandfathers passed on while I was a teenager, but my grandmothers live in our hometown. Both have played a huge part in my children's lives. The children's great grandmothers regularly cared for them when I was working or at college. As my children and their great-grandmothers got older, it was more of a mutual day care. One of their great-grandmothers passed on just two years prior to the children's graduation, a few days shy of her ninety-first birthday. She was still interacting with them daily. My children had the same strong, loving family roots I had. That's what family is all about: loving, caring, and sharing on a daily basis. I am blessed and fortunate to have such a family, and I thank God for them! My family has been a tremendous asset and foundation from which our strength and security abound! My children grew up in the same neighborhood I did. Many of our neighbors have been long time friends of our family for generations, playing a part in our lives.

My in-laws helped us build our house, which we moved into a week after Drew's birth. We will be forever grateful to them for this. Our home was in the semi-finished state when we moved in. The basement had the basics and the upstairs was an unfinished shell. The plan was I would finish the rest of the house over time. I did accomplish quite a bit, but as money became tighter and tighter due to this disorder, it became harder and harder. To make a long story short, I never finished before ordinary upkeep and repairs set in. Many people know if you have the time you don't have the money. If you have the money, you don't have the time. Sometimes you are without either.

I was brought up a Catholic. My dad's mom was very religious and she constantly reminded us of our faith. I can't say that I am a devoted Catholic. We no longer attend church regularly for a number of reasons, but I remind my children often to talk to God and thank him for what we do have. Despite its being such a struggle to hold onto what we have, I consider myself very blessed! Material things are not the most important or a measure of true value! It's the family, friends, and all the rest of the treasured individuals God has blessed us with that make the difference in our lives. My faith and belief in God has only become stronger through our struggles. I don't want to preach to anyone about God. I like to think, that despite all of the differences and beliefs about God, we are all praying to the same God, just in our own unique ways. This is just my opinion. Religion is a very personal thing between a person and God.

The following experiences that I feel were related to God are important for me to put into my book, but each and every one of you reading this will draw your own conclusions. I have always talked to God regularly, not only to thank him but to ask for strength and support. Sometimes I would cry myself to sleep after watching

Randi cry herself to sleep, suffering in such pain day after day and night after night. I have even contemplated what I would do to end this horrible suffering for my children. I know you may think I must be crazy to think this way, but I was so desperate. None of their doctors could or would do anything to ease the daily overwhelming pain and suffering. It seemed likely each one of my children would only continue their downward spiral to their deaths. I would rather all of us go peacefully together. I wondered if possibly Drew, the least affected, would receive help eventually before he reached the same hopeless point as his sisters. I was apprehensive about including him in such a drastic measure. I don't know if I could have left him behind knowing what he would have to face all alone. I don't know at what point I would have made such a decision or if I would have done it at all. I am glad it didn't come down to that!

Randi was twelve years old when she started Neocate. She never slept through a night prior to this because of multiple pains. On at least two occasions, while I was talking with God, I felt the thoughts I had were more than just my own personal thoughts. At some point in time before Neocate, when I was watching horrible amounts of ever increasing pain and suffering daily, I asked God,

"Why do my children have to suffer so much?" I didn't see any faces or anything. I didn't hear any voices per se; I just had these extremely strong thoughts. I questioned myself, "Were these just wishful dreams of how I hoped things might turn out?" I just knew there was an intense feeling attached that made these thoughts stand out and different from the ordinary. I was encouraged to somehow prove beyond any doubt that these strong thoughts were real and would become reality. I thought about writing it all down and sealing it away in an envelope or to let others know. This would be proof that these were God's actions. When I asked God what I had to do to repay him, the suggestion was just to tell this story. Well, I certainly had enough people, doctors mainly, who already thought I was crazy. I needed to be careful who I would share this with. First and foremost, I shared these things with my children. I wanted them to know that God was with us, and I wanted them to know of his works. My son tried to explain it to our pediatrician one day when we were in for a visit. Luckily, she said she couldn't understand him. However, she did catch the part of it having something to do with God.

She said to Drew, "Don't you know that moms have a direct line to God?"

Besides my children, I told a few other friends who were a big part of our lives. They knew me and knew I was not crazy. Those strong feelings were right. All of the things became reality. I couldn't have made them happen. The first thought was that the children's pain and suffering would end soon. I questioned what was considered "soon." Would it be a few days, weeks, what? However, there were no further strong suggestions or thoughts. It wasn't like a question-and-answer interaction, unfortunately. The next thought was: the next place you go, you will find the answers and help. I asked how to choose the next place, "Surely, you must need to give me some kind of further sign or direction. How will I know where to choose as the next place to go?" I said to myself, "You can't tell me that just any place I choose will bring us what we've been searching for." You can see that, even though I had these intense feelings, I was very skeptical. The last revelation, or whatever you want to call it, was there would be one thing, a medication of some sort, which would resolve all the children's pain and suffering and would replace all the medications they were now taking. Well, you've read our story. I no longer fear death, or look at it the same any more. I have an unquestionable belief in God which I find very comforting.

The day I found out all of the doctors from B.CH.H. had given up on my children, I was terribly traumatized. Once again, God was there for me. This time, after hours of crying and asking God for his guidance, I experienced this instantaneous feeling and vision of a white, wide wingspan that gently wrapped around me and embraced me like a hug. In the morning, when I awoke, I remembered this was the last thing I recalled before falling asleep. I thought to myself, "This was God's doing."

It was God's way of soothing me and taking me away from the pain I was going through. I wondered, "Is this part of or similar to what people experience when they die, this instantaneous peace?"

THANK YOU, GOD!

ACKNOWLEDGMENTS

This is my last big batch of lemonade! I could not have even thought of undertaking this project at this time in my life if I didn't presumptively assume I automatically had the support and help of my faithful, trusted, loving family and dear friends. THANK YOU ALL, for making this possible! A special thanks to Kim, Thelma and AML, "For helping me with you know, whatever, this and that thing to make it more better." I love you guys! I can't forget my computer man! I've found it's not what you're faced with that matters as much as whom you have to help you face it. I don't claim to be a writer. I have no aspirations of becoming one either. All I know is I wanted and needed to share our story.

Approximately ninety percent of my book had been complete for several years. Through the years my desire and commitment to finish it has been on and off again. I wanted and needed to wait until we had a definitive diagnosis to finish it. The driving force behind embarking on this endeavor presently is the need to eliminate the many obstacles, struggles, and extreme frustrations which remain for individuals and families afflicted with mitochondrial diseases.

Medicine is an ever-evolving changing science. There are currently varying opinions and views regarding many of the aspects surrounding mitochondrial diseases. I have reiterated some of the current known facts in my book. For more information, I have included a resources page.

ADVOCACY TIPS

The following is a checklist of advocacy tips. I prepared it specifically to deal with insurance companies. It can be applied as a guide in many situations.

- ✓ 1. - Always follow up any telephone conversation in writing. Make sure the correspondences are dated and sent by certified mail.

- ✓ 2. – Keep track of your phone calls. Jot down notes regarding the conversations and save your phone bills for reference.

- ✓ 3. – Keep good records. You can use them in court. As time passes and things become more involved, you may not always remember little things that may be important.

- ✓ 4. – If the insurance company denies you coverage, yet you feel it's wrong, don't be intimidated. Stand your ground. This wasn't my first battle. Since my children's birth, I've had to fight for proper medical care, treatments, schooling, services, etc. Usually initially, you're told "no" or denied coverage / services. You have to dig deeper and fight. Your perseverance will pay off.

✓ 5. – Have others who know you and your circumstances, e.g., doctors, nurses, teachers, therapists, etc., write letters of support on your behalf.

✓ 6. – Contact magazines, T.V. – both local and national, support groups, local agencies, government officials. Everyone and anyone you can think of that might be able to lend you assistance. Follow up on all leads. You never know where they may take you.

✓ 7. – KEEP TRYING! It will take a lot of time and patience. At times, it may seem overwhelming and frustrating. Stay focused and remember why and who you are doing this for.

COMMITTEE ON SPECIAL EDUCATION (CSE) ISSUES

Our pediatrician felt an outline for parents on what to expect from the Committee on Special Education (CSE) meetings and how to work with your school system to get the most for your child academically would be of value. This process can be very frustrating and time consuming. We were one of the few families to establish a good working adaptive school program for my daughters and were often used as a model for others.

C.S.E. meetings are mandated by law as a forum in which the school and the family / guardian(s) get together to develop a plan to meet a child's special needs. C.S.E. meetings are generally held annually. However, you can request a C.S.E. meeting anytime. Triennial reviews, as the name implies, are held every three years. The in depth reassessment of your child includes updated tests and a recent psychological evaluation. The committee is comprised of district staff, a parent advocate, and parent(s) / guardian(s) as well as others invited by the district or parent(s) / guardian(s). Many parents may be intimidated by the thought of meeting with a group of individuals they may not know, but it is important to attend these meetings.

First, the child will need to be identified as a special needs student. Quite often this will be the first battle parents must fight. Some parents trust the school will do the best thing for their child. Although the purpose of these meetings is to make decisions that are right for your child, not everyone always agrees as to what that is! Often the schools are more focused on their money and their convenience. Your child shouldn't be warehoused till age twenty-one, but this is often the preferred and easiest solution for the schools. As parents of children with mitochondrial disease, you must educate your CSE about their unique needs. Parents know their child better than anyone else and are the best advocates when attending these meetings. It is important to know your rights and protect the best interest of your child. The priority is having all of your child's safety and medical needs addressed appropriately by a competent staff while in the school setting. I was surprised to learn many parents of special needs children don't get involved and never even attend a C.S.E. meeting though they must be invited by law.

It was not easy for our family. We had to fight for everything we got! That's the key whether it be the school, insurance companies, Medicare, Medicaid or doctors. You must be persistent and take control. As with most investments, the more input, the more output. There was trouble from day one with our school district but through persistence we were able to succeed. Education empowers an individual! In my opinion, our schools can never do too much for a child because ultimately it will benefit society. Here is a list of helpful tips that worked for us.

By contacting any one of the resources listed below you will be guided to the others.

- Find advocacy groups in your area that hold seminars on mainstreaming and advocacy for special needs children. Attend their meetings and educate yourself so you can stand up for your child's rights. There will be printed information available at the meetings. Talk with some of the key speakers, if possible. Sometimes they are willing to personally get involved, as was the case for us. Ask how to get free help and where to find it. Usually there are lawyers specifically dedicated to this line of work that work for free or a reduced fee in clinics.

- Find a private neuropsychologist you and your child like.

One you feel understands your child's needs. The school will have their own psychologist and your child will be evaluated by this person. However, if you don't agree with the report or the recommendations, the law is on your side. You are entitled to an independent evaluation by a doctor of your choice. The C.S.E. is not going to want to inform you of this or pay for it, but it's the law! A good neuropsychologist can make all the difference. The doctor will compile a report with specific recommendations for the school to implement after determining exactly what your child's weaknesses stem from. The school will not necessarily be willing to follow all of the recommendations. Again, it will be up to you to ensure this happens. You are the key! You must remain involved. In our case we were not going to wait for the school to drag the process out while an entire school year was wasted. So, I borrowed money to get the evaluations done and sought reimbursement from the school district later.

- We had to enlist the help of lawyers from our local law clinics

a number of times to enforce or obtain what was needed for my children. This all takes time but after a while the school respects you and knows you mean business and that you're willing and capable of doing what it takes. At one point in an effort to continue providing my children with an education, while we were battling, I hired a private teacher for which I sought and later recovered reimbursement from the school.

- If the psychologist recommends a full time aide, fight for it

with everything you've got because, if you find the right person, that individual can make all the difference in the world! Daily one on one nurturing by a competent, caring individual will go a long way to your child achieving all they can. Safety issues and medical concerns will be handled by this person. An aide also becomes a valuable liaison between you and the district. Follow-up with the neuropsychologist at least every three years when the traditional triennial reviews are mandated by law. ***Continuity is a very powerful tool!***

- Document, document, document, everything! Write follow

up letters referencing all conversations or outstanding issues. Make sure you have everything in writing, and keep copies. Send correspondences by certified mail. Stay organized! Ask for copies of all school generated reports. I have drawers full of files and files of paperwork which I've needed from time to time. Obtain documentation from other doctors involved in your child's care especially if medical needs are involved. Because of such documentation, our school was obligated to provide at home tutoring and an on campus isolated classroom for my daughters.

- Audio tape the meetings. This puts an end to the nonsense of the school personnel not having the same recollection of the issues presented or promises made at a previous meetings. You do have this right! In N.Y. State taping meetings was just one of many things our family with the help of lawyers fought for and obtained.

- Bring advocates with you to the C.S.E. meetings. They can be family, friends, or anyone you feel can help you. Even if it's just to level the playing field with several school personnel present. It's easier for you to get things accomplished when you have witnesses and support in these proceedings. It's not just your word against theirs. As time goes on, you should gain more help and support of others involved with your child's care. They will be valuable in obtaining what is in the best interest for your child. In our case, it initially was the girl's teacher of the visually impaired who recognized and understood my daughters multiple needs and was willing to advocate for them. She would be present at every C.S.E. meeting or we didn't hold the meetings until she could attend. Then it was the personal aides who would advocate on the girls' behalf even though they were school personnel.

- Stay in direct contact with the school personnel. Know who's involved and what is happening, always! Don't assume everything is O.K. because your child is at school and in the hands of those you feel you can trust. The more involved you are the more you'll realize how important it is to be involved. I always brought my children to school and picked them up. I chatted with the classroom teachers. In the girls' earlier years, we used daily communication notebooks. I would write daily concerns and problems in it. In return, the school personnel would do the same. Go on field trips with the class. Get to know the other children in your child's class.

- Ask to hold regular meetings so everyone involved voices concerns before major issues arise. This helps keep the line of communication open between you and the school personnel. We called our meetings, transdisciplinary meetings. These were held whenever necessary but at least monthly as long as the girls were in school.

It was a big struggle throughout my daughters' school years but in the end I'd have to say, "It was definitely worth it." The very people who doubted my daughters' medical problems and needs for modifications grew to respect me and cared about my children. In fact, the C.S.E. chairperson and his family were invited to my children's graduation party and they attended. My girls' accomplishments were more than anyone could have ever imagined!

A word of caution regarding requesting an impartial due process special education hearing: a recent Supreme Court decision, _Schaffer v Weast 2005_ overturned a policy that existed in the special education field. Previously when a hearing was requested by a parent the district had the burden of proof. This was because the district had more information, personnel, and legal resources. Because of this ruling, it is now the plaintiff who must bear the burden of proof. This is usually the parent. Most parents who request an impartial hearing do not have the financial or legal resources to advocate effectively for their child. In some cases, districts are using this decision to intimidate parents by informing them that if they disagree with the district over what services are appropriate for their child, they have the right to a hearing and they can either hire an attorney or act as their own. Check with your state legislators to determine if they are advocating to move the burden of proof back to districts either through law or regulation; if not urge them to do so!

RESOURCES

These are just a few of the resources available. By contacting any of these you will find many more.

M.D.A.C., (Mitochondrial Disorder Action Committee), can be found on the web at www.mitoaction.org. Address M.D.A.C. 14 Pembroke Street, Medford MA 02155-4827.

The United Mitochondrial Disease Foundation (UMDF)
Promotes research for cure & treatment of mitochondrial disorders.
New England Chapter of UMDF 39 Bay Farm Drive, Plymouth, MA 02360
Family support: Beverly Ingram 413-593-5920
http://www.umdf.org/

National Office: 412-793-8077

The Muscular Dystrophy Association (MDA)
The MDA offers a vast array of services to help you and your family deal with mitochondrial disorders.
http://www.mdausa.org/
National Office: 800-572-1717

Rebecca's Guide, Inc.
A website that provides information about mitochondrial disorders and the people, activities, and organizations involved.
www.mitoinfo.org

The Genesis Fund
A nonprofit that raises money for the specialized care and treatment of New England area children born with birth defects, mental retardation and genetic diseases. A major goal is to provide state-of-the-art, coordinated, humanistic care to these patients.
www.thegenesisfund.org

Marcel's Way
A non-profit organization established to provide lives touched by Mitochondrial Disorders with information, education and support.
www.marcelsway.org

The Neuromuscular Disease Center at Washington University in St. Louis, MO provides a technical understanding of mitochondrial disorders. www.neuro.wustl.edu/neuromuscular/mitosyn.html

Wish Upon A Cure www.wishuponacure.org
An organization dedicated to raising awareness of mitochondrial disease. Is currently funding the first ever US clinical fellowship in Mitochondrial Medicine at the Mitochondrial and Metabolic Disease Center (MMDC) at the University of California, San Diego School of Medicine.

National Organization for Rare Disorders
The National Organization for Rare Disorders (NORD), a 501(c)3 organization, is a unique federation of voluntary health organizations dedicated to helping people with rare "orphan" diseases and assisting the organizations that serve them.

EPILOGUE

THE BIG CIRCLE

I'll be the first to admit anything is possible at any time with mito, but what has transpired since I finished my manuscript a year ago astounds even me. When none of the doctors in Philadelphia could make any further allowances for my young adult children due to their ages, we reluctantly agreed to have them see the adult neurologist I was referred to in Philly. Both Randi and Brooke had several new symptoms and ever worsening medical problems we needed to address sooner than later. Randi's pediatric doctors felt she needed hospitalization to evaluate her current symptoms and problems, but we had no adult doctor that could admit. It was a several months' wait before the girls could see an adult doctor. As is standard practice, the first appointments involved evaluations and review of past medical histories. It would be another four or five months before we could get the next appointment to actually begin addressing the girls' problems so they could receive the help they needed. I had an appointment of my own, for which I had waited two years, upcoming with this doctor. I wanted to give it to one of the girls, but I couldn't decide if Randi or Brooke needed it more. When I informed the doctor's office of my plan, they told me I could not do this. In return, out of extreme frustration, I said, "I guess we need to find another doctor, one who has the time for and cares about their patients."

Despite making the trip to Philly for this appointment, none of us were allowed to see this doctor. That was fine with me! I did NOT want a doctor who could not understand why a mom would do this because of all the suffering her children were going through while they waited and waited to see the only doctor that could help them. I felt an urgency for doctors to do something to stop the suffering and deterioration I was once again witnessing in my children. What parent wouldn't do the same thing under the same circumstances? In the past I had to standby and watch while no one did anything because no one knew what we were dealing with. Now it was a different story. We had a diagnosis and still no one was doing anything. I swore I'd never let that happen again! I realized my stubbornness left us with no doctor that could help us and we'd wasted several more months' time. Not having any idea of what to do next, I kept in mind things can work out even when it seems impossible. One of the things this doctor had done during the initial evaluation was to refer us to a metabolism specialist in Boston, of all places, to address the children's malabsorption issues.

Several more months elapsed before we could get an appointment with the doctor in Boston. Actually, it was six months to get one girl in and it would have been eight months if I wanted to have them both seen at the same time. Once again I had to choose which girl I felt needed help sooner. Past experiences had proven what was found in one would help us all. Time was the biggest factor so the sooner we got one in the better.

Meanwhile, our pediatrician called one day to say she was going into semi-retirement, and the rest of the practice didn't feel comfortable seeing Randi, Brooke, and Drew since we had almost exclusively seen her for the past twenty-four years. She had forewarned me but I dreaded the day I would hear these words! This came on the heels of another phone call from our pediatrician informing me she was concerned Randi had significant heart problems based on a recent E.K.G. She had consulted with a cardiologist and I was told Randi had global myocardial dysfunction and left ventrical enlargement. Speechless, I couldn't believe what I was hearing. Randi needed to be evaluated by a cardiologist without delay. Having Randi seen by a cardiologist familiar with mitochondrial diseases was best, but our pediatrician felt we may not have the time to wait and she alone couldn't handle heart failure which she felt was imminent for Randi.

We needed to find another local doctor. Our pediatrician agreed to consult with prospective physicians for us. Many times in the past our topic of discussion was how several doctors contacted all refused to take on our cases. I told her she could try but I didn't think we'd find anyone willing to take us on. In fact, the next doctor I suggested we try also refused when our pediatrician called her. However, she had just gotten a new partner, and she was willing to take our cases. I've learned there are two types of doctors: the first group refuses because they know nothing about mitochondrial diseases; the second group agrees to take on the cases only to later realize they shouldn't have because they know nothing about mitochondrial diseases. Reluctantly, we agreed to give the new local physician a try, but it would be months before we could get an appointment. Admittedly, our pediatrician thought it might take several years and several different doctors before we found the right one. This was the last thing we needed right now! It was bad enough that we had to start all over from square one with all new specialists. Now, we had no local doctor. We had no one at all! Yes, I've said, "I wish we didn't have to have any doctors," but unfortunately that's impossible given our circumstances.

This was all too much! How do you bring a doctor up to speed on a condition you've been dealing with for the past twenty-four years when the doctor doesn't know or understand anything about it or you? This was an overwhelmingly daunting task, one I didn't feel like tackling! I was upset and on the defensive. I was opposed to taking several giant steps backwards after all we'd done to get where we were! It was May and Randi's appointment with the metabolism doctor wasn't until August, the earliest we might have one doctor that could do anything. We'd just have to go to the E.R. if something came up. There weren't any other options. I can count on one hand the number of times in my children's lives we ever had to go to the E.R. It wasn't something I looked forward to.

In my efforts to find someone to help me with the publishing aspects of my book, I searched the internet for others who were on the same mission. With the law passage experience, I learned

aligning with others with the same cause can be very helpful for

everyone. By luck only did I find M.D.A.C., the Mitochondrial

Disease Action Committee, based in Boston. I contacted the group

to see if they might be interested in or able to help me with getting

my book published. While exploring their website further, I found

the metabolism doctor we had been referred to, whom Randi had

an appointment to see in a few months, was on the board of

M.D.A.C. Reading on, this doctor was described as the primary

mitochondrial specialist in Boston. Whoa, this was exactly what we

needed! Once again, things do happen for a reason after all! My

level of anxiety and fear was extremely high, however. What if he

told us he didn't think we had mitochondrial disease? What if he

was like all of the other doctors in Boston in the past? What if this,

and what if that? I was a nervous wreck! We'd just have to wait

and see.

August finally came. It was time for Randi's long awaited

appointment. The mixed bag of anticipation was nearly over.

Since we would need to be traveling the day before our

appointment, we wouldn't be home to receive the confirmation call from the physician's office. I figured I had better call and confirm

our appointment with the doctors office. When I did, I was told

Randi didn't have an appointment for the next day. I was livid! I

immediately demanded the person I was speaking with find

someone who knew what was going on and could help me because

we planned on showing up for the appointment no matter what!

Shortly, I received a follow-up call from the physician's office. They

had found their error and told us to come for the appointment.

Why, oh why ,oh why, do things like this have to happen? To my

surprise the visit was pleasant. Addressing Randi's ever worsening

problems, now well over a year and a half old, would have to wait

until we established care with a whole new set of specialists in

Boston. The majority of these new specialists were also

pediatricians, the very reason we had to leave the last group of

specialists. In Boston, they came up with varied ways of getting

around this obstacle and made allowances.   Whereas, in Philly,

they couldn't. With every new specialist the girls and I met, I was

haunted by the past trauma so deeply ingrained in me. Over time I

realized I could lay this anxiety to rest. Every specialist we saw

was part of a network of physicians, all familiar with mitochondrial diseases. A great deal had changed in Boston over the past twelve years! Even in such a cutting edge medical mega as Boston, doctors tell us the majority of their colleagues still know little if anything about mitochondrial diseases. This very proactive network is on a mission to change this. Our family is extremely appreciative of all the hard work parents, doctors, nurses and other individuals have put into this exceptional network and of their continued efforts to improve the care for mitochondrial patients. We were able to walk into this network and reap the benefits. It's a welcomed stark contrast to what we've experienced in the past. Forearm stress testing, which Drew had undergone previously, was ordered by our new doctor for the girls and myself in an effort to establish further documentation for mitochondrial dysfunction. Similar findings in Brooke and me pointed to mitochondrial dysfunction because our muscles weren't utilizing oxygen properly.

"What a big circle," were the words used by our pediatrician when I updated her since we had last seen her several months earlier.

Boston was the last place anyone would have ever imagined we'd end up! It was the very place I swore I'd never return to, no matter what, since it was the place that had given up on my children twelve years ago.

Shockingly, we had documented in October, Randi's dysmotility had returned. The wavelike muscular contractions that propel the food through the gastrointestinal tract were not working again. She was and had been for over a year unable to take anything by mouth without extreme nausea and at times vomiting accompanied by constant stomach pains. Her gastroesophageal reflux was no longer being controlled by medications. She was also beginning to complain of trouble swallowing. These gastrointestinal problems were neither due to peptic ulcer disease, a previous G.I. doctor had diagnosed, nor were Randi's G.I. problems related to a viral infection, as I suspected. How could this be? What had happened?

Randi, her brother, and sister had been free of every gastrointestinal problem for the past ten years. Gastroesophageal reflux had returned in all three over the past two years. Frantically, I tried to make a connection but there seemed to be no common thread. They were all on Neocate, the very savior that had gotten them out of this mess twelve years earlier. With Neocate no longer alleviating their multiple gastrointestinal problems, we were right back to the place where we were twelve years ago. The list of medications all three had been on for several years prior to Neocate, Randi was again prescribed. Knowing these medications hadn't worked in the past, I was very doubtful they would be useful now. We'd been there before and done that!

Reglan, a popular dysmotility drug, was reintroduced. As a child Randi seemed to tolerate Reglan, whereas it had caused parkinsonism in Brooke. A startling discovery was made. Reglan not only caused an increase in Randi's baseline tremor and twitching, but it affected her mentally. She felt very agitated, nervous, and complained of significant depression. In retrospect Randi feels she had similar feelings as a child but was unable to convey this to us at that age. Although Reglan helped the dysmotility to some degree, it had to be stopped due to the side effects. Randi's condition continued to deteriorate. By January, she wasn't getting more than a few ounces of fluid intake daily. The only option left was to have a feeding tube placed.

An esophagoscopy and manometry study were scheduled to evaluate Randi for a feeding tube into her jejunum, the second part of her small intestine. Originally the plan was for Randi to leave the hospital post the procedures with a temporary N-J tube in order to receive nutrition and hydration. Before surgery to place a permanent J-Tube, we needed to assess whether or not Randi tolerated the jejunum feedings.

During the esophagoscopy, direct inspection of the gastrointestinal tract, the G.I. doctor found Randi's villi to be flattened and the mucosa looked smooth and shiny. This indicated likely disease of the small intestine. Villi are tiny finger like projections that are responsible for the absorption of nutrients from the foods we ingest. The physician stated we could not use the small intestines for feedings if there was damage. Biopsy results wouldn't be back for at least one week. Only then would we know how we were going to feed Randi. I could not believe this. Just one year ago, almost to the day, Randi had her last esophagoscopy.   Specific attention was paid to the villi because of the issues of malabsorption we felt all three were complaining of or exhibiting. A pathologist review showed the villi were fine. The new G.I doctor said he'd be very surprised if the biopsy results didn't show celiac. Celiac is an inherited disease caused by ingesting a protein called gluten which is found in wheat, barley, rye and possibly oats. This sets off an autoimmune response which damages the small intestine's villi, in turn causing malabsorption, malnutrition and a host of other related symptoms. If celiac was not found, the specimen would be tested for other things that could cause the villi to look this way. He seemed confident there was something wrong. Although surprised with these results, I was relieved because this meant we were finally going to get to the bottom of the looming malabsoption issue. The question was how was I supposed to keep Randi going for the next week while we awaited the biopsy results since she didn't have the temporary feeding tube? "Just keep trying to get fluids in," I was told.

Upon returning home from Boston, Randi was up all night vomiting after only consuming approximately six ounces over the past twenty-four hour period, and she couldn't take her medications. All night I lay awake pondering what I was going to do. My mom and dad had just left for a vacation in Florida, and I was to care for their home. I didn't want to tell my mom and dad Randi was in the hospital because I knew they'd come home, and they had just arrived in Florida after a three day drive. Brooke needed constant care and Drew was away at college. The extreme

fatigue I was experiencing made driving back to Boston dangerous.

When I get to this level of exhaustion my vision fails and my brain

shuts down.

By morning Brooke and I had come to the same conclusion.

Bags were packed and we headed to our local E.R. Brooke would

have to stay with my brother and his family if Randi needed to be

admitted. I would run home when I could to check on the houses

and feed the cat. The E.R. doctor confirmed my suspicions. Randi

was in bad shape and needed to be hospitalized. She was

severely dehydrated and in metabolic acidosis.

The hospitalist who was placed in charge of Randi's care

since we had no local doctor stood before me saying, "Why didn't

you take her back to Boston? If it was my child in this situation I'd

take her to Boston."

I knew Boston was best, but under all of the circumstances

at the time, it wasn't possible. I didn't think the hospitalization was

going to end up being as long or involved as it was. I envisioned

arranging with the E.R. doctor to have our home health care

agency provide hydration intravenously at home for Randi while we awaited the biopsy results. On my request, along with the hospitalist's own desire to come up with a plan, he placed calls to our mitochondrial doctor as well as our G.I. doctor in Boston.

He returned and once again stood before Randi and me with arms folded and said, "Your doctors in Boston can't admit Randi because they're both pediatricians."

We were warned from the start that if a hospitalization were needed it could be done but it would be a complicated process. This was outrageous and a perfect example of how poor the care for mitochondrial patients remains despite having doctors involved familiar with mito! It was overwhelmingly frustrating to be abandoned at a time you need help the most! The hospitalist reiterated that regardless of whether or not either of our physicians in Boston could admit Randi, I should get in my car and drive her to Boston right now. Circumstances did not allow me to do this. In my overwrought state I felt hurt and didn't want to be someplace we were not wanted. I knew this wasn't really the case; rather it was the dysfunctional system but nonetheless it was upsetting.

At our local hospital the hospitalist and the rest of the staff did their best and Randi received good care. However, months later we are still dealing with the consequences of decisions made by a doctor who knew nothing about mitochondrial diseases. Many of Randi's medications were not administered. I tried explaining and eventually arguing each issue, but every decision made was ultimately the doctor's. Randi was placed on intravenous fluids while a plan was devised. T.P.N., total parenteral nutrition, was the only choice. T.P.N. is used when a patient's gastrointestinal tract can't be used. Randi would require a PIC line, a special long intravenous catheter, which is placed into a deep major vein leading directly to the heart. The problem was we had been warned by several doctors in the past to avoid such devices because of her immune deficiencies. Again I wanted to consult with a specialist, but we couldn't. We hadn't established with a new immunologist in Boston yet, and our mito doctor was out of town and not available. Our local hospital's standard formulation of TPN was not safe for Randi. It contained proteins known to be intolerant for Randi. I had to wing it and hope I had enough knowledge to make the right decisions. We were on our own once again.

After tolerating T.P.N. for a few days, Randi was set to be discharged home where she would continue to receive TPN . We were very accustomed to and comfortable with home health care which we preferred over hospitalizations. For years the girls had received intravenous medications at home. Documentation from our G.I. doctor for insurance to pay for the TPN was all we were waiting for. Instead I received a phone call from the G.I. doctor telling me the villi looked functional under the microscope, and everything tested for was negative. Therefore we could now use Randi's small intestines for feedings. Randi hadn't left the hospital although she'd been discharged. She was readmitted in the computer and had a temporary N-J tube placed. Randi began weaning from TPN while at the same time feeds were introduced through the temporary N-J tube. In the meantime, my mom and dad figured something was wrong when they couldn't reach me at home a few days in a row and, as expected, they headed home to help.

After ten days Randi was finally discharged with the temporary N-J tube. The plan was to make sure Randi tolerated the jejunum feedings before we put her through the surgery to have a permanent J-tube placed. After only a few days, the best case scenario started unraveling when the temporary N-J tube began malfunctioning. A great deal of backwards pressure was coming from inside of Randi or at the end of the tube placement , or something. No one knew what was wrong. A pressure relief valve and bag set were sent to us after consulting with our home health care agency. For approximately twenty-four hours this seemed to be working. No one informed us this might happen or what we needed to do if it did. This was deja vu all over again! This is how one gains lots and lots of hands on experience the hard way. Randi now required constant care. Of the two girls, Randi had been the one who could self medicate. Now everything had to be in liquid form measured out and gravity fed through a syringe. It became a nonstop job because of all of the medications Randi required.

Everyone kept telling us the permanent tube would be much better, and we shouldn't have all of these problems. I hoped this would be the case, but how do you ever know for sure? I left Randi for about 15 minutes to lie down and rest. When I returned, Randi was complaining of refluxing and being sick to her stomach. She thought there was something wrong with the tube. The temporary tube had a mind of its own. It made an unexpected exit from Randi's small intestines despite being taped perfectly in place and not disturbed in any way. I quickly noticed it had pushed itself beyond the clamp still taped in place properly at the end of her nose.

In a few minutes I was supposed to take Brooke out with her boyfriend on a date for Valentines Day. My mom was supposed to watch Randi during this time. I called my mom and said, "There's been a change in plans. I need to take Randi to the hospital. Can you take Brooke on her date?"

When you are totally dependent on a tube for all nutrition and hydration, there are no other options if something goes awry. Back to the E.R. we went. After repositioning the temporary tube and two more hospitalizations locally that was it! We wanted the permanent tube placed. I was warned that because we had not conducted a long enough trial with the temporary N-J tube, there was no guarantee Randi could tolerate the Jejunum feedings. She may ultimately end up back on TPN. I was told this sometimes happens. Then parents are very upset that doctors put their children through surgery. No one involved had a crystal ball. It was one of those times when I had to take a leap of faith and hope and pray for the best. I was sick with the worry of making a wrong decision, but circumstances dictated I must make a decision.

Never having driven directly to Massachusetts General Hospital, I was anxious about attempting it, but I was too tired to waste precious energy fretting about the drive. In earlier years our medical trips were in a different section of Boston. When we left home the temperature outside was below zero. Randi had an I.V. line, and we had a pump and fluids to cart around. Randi was barely standing say nothing of walking. I won't go into all of the details of the trials and tribulations we encountered just getting to Boston, but I am sure you get the picture. It was a lot to take in all at once, but I couldn't think twice about it since I knew it was best for Randi. While conducting preoperative testing, there was an abnormality found on Randi's EKG which raised concerns, but we assumed it was just the same abnormality we had noted for the first time several months ago. A few weeks earlier, Randi had a cardiological evaluation. We were then told the EKG findings were nothing to be concerned about. We had no reason to anticipate any problems.   We just prayed for the best.   Only an overnight stay was customary after this procedure, we were told.

I didn't think it was going to be as rough as it was on Randi. Despite Randi and I being introduced to the anesthesiologist assigned to her procedure when we arrived at the O.R., we were met by a different anesthesiologist. It was the anesthesiologist who had administered anesthesia to Randi a few weeks earlier. He announced he'd be taking over and immediately asked for a pediatric intensive care bed to be placed on standby. I was told it was just a precautionary measure and if anything went wrong he would stop the surgery and send Randi to the PICU. This seemed strange to me on the one hand, but on the other I felt a little more at ease knowing this doctor knew her better than someone new.   I was told to go back to Randi's room and wait for the doctor to call. Two hours later the call finally came. Everything was fine, I was told, and I could go see Randi in recovery. However, I was also told we were very very lucky this time, but Randi should not undergo any more surgery unless it was absolutely necessary. "Well, that's just common sense," I thought to myself. However, this was reiterated over and over by a number of the doctors involved throughout Randi's admission. Concerns were raised that Randi wouldn't be able to breathe on her own post surgery and would need a respirator. No one would tell me exactly why all of these concerns were raised or what they were based on. I do know Randi was in a good deal of pain despite the administering of an adequate amount of morphine. Unfortunately, we hadn't found any pain medication that helped Randi to any significant degree. She was nauseous, which is to be expected, and in severe pain.

She was arching her back and writhing from the table saying, "I can't breathe."

I could see by the monitor she was receiving adequate oxygen. I had just experienced a similar episode myself when my diaphragm muscles spasmed while pacing the floor waiting for Randi to come out of surgery. It is a common but very painful and limiting symptom we both share, so I tried to calm her, assuming it was indeed just her muscles.

While attempting to comfort Randi, I began feeling very weak. I hadn't gotten a chance to lie down all day and nearly passed out. I knelt down and lowered my head for several minutes. I was unable to address my exertional fatigue in this situation.

For weeks Randi had a great deal of increased muscular cramping, spasming, and the associated pain post the surgery. It hurt her to breathe, to urinate, and to move. She had her usual baseline pains and other symptoms on top of all of this. Some questioned why she wasn't bouncing back and discharged already, but like the doctors who have studied mitochondrial disease, I knew this extended recovery was very common.

Days later it was felt nothing more could be done in the hospital that couldn't be done at home. It was believed she would benefit from being in the home environment where she was more comfortable and could get more rest. I couldn't wait to get home! I'd had more than enough of hospitals over the past month! I knew it was just going to take time, lots and lots of time, based upon Randi's past history. She doesn't recover from anything in a normal amount of time. Leaving the hospital, I dreaded the thought of having to return in little over a week for a medical trip planned months earlier. If it weren't for the fact that we had waited several months for these much needed appointments, I would have canceled the trip.

At least there were no problems with the new tube. It was functioning very well. We had enough other problems to keep us busy around the clock. About a month prior to her hospitalizations, Randi complained of all her neuropathies worsening, and she was falling more. Growing ankle instability was a problem. I attributed the increased tripping to this. Chronic back pain became more severe as well. We had managed to get an MRI of Randi's L spine along with an EMG of her lower extremities in between all of the hospitalizations. Randi had significant nerve damage to her common peroneal nerve at the fibula head. Either she compressed the nerve against something or it was from her significant weight loss. It would be nearly a year, if no further injury occurred, before we could expect a partial recovery for semi-normal use of her right leg and foot. This meant someone had to assist Randi every time she needed to go to the bathroom or to and from her bedroom. She was given a script for a walker, but she was too weak, unsteady, and lightheaded to maneuver it on her own. I wasn't taking any chances on the J-tube being pulled out! We didn't want more surgery or problems in any shape or form. Randi was still having great difficulty sitting up or standing. She was very dizzy, felt like she was going to pass out, and had associated sweats/chills, tremoring, nausea, and at times vomiting. Dysautononmia was the culprit. It also explained Randi's abnormal EKG's, increasing tachycardia and palpitations. Thankfully she does not have heart damage! Her sinus infection took a good hold because of our inability to medicate Randi properly in the past weeks.

Randi was a mess! Time and only time would tell what Randi's new baseline would be. Randi went from being pretty much self-sufficient for the basics, in the home environment anyhow, to needing constant care. Life as we knew it was forever changed again, this time not for the better. When Randi was able to tolerate it, she was fitted for new braces. We began using a wheelchair as we did years ago.

Although it was a major ordeal, the tube feeds alleviated the GERD and the extreme chronic nausea and vomiting that was associated with all oral intake. We now had the ability to get nutrition and hydration into Randi. Apart from the very time-consuming tube feedings, the hassle of medicating, the awkwardness of being tethered to a tube and bag that needs to be toted around which Randi can't carry on her own, it's an improvement. Thank goodness such alternatives exist. I shouldn't complain at all! Another good thing to come from all this was we finally found a pain medication for the first time in Randi's life that is effective. Although the morphine she was administered post surgery did little for Randi, she was sent home on oxycodone, which to everyone's surprise, worked. We now have a means of helping Randi with her chronic ever-increasing multiple pains. A sleep study to evaluate Randi's breathing confirmed what we already knew; Randi can't sleep mainly because of pain.

Will Brooke and Drew follow in Randi's footsteps, as is usually the case? Brooke and Drew have their own issues we are currently addressing but they pale in comparison to Randi's problems at this time. Brooke's chronic allergies continue to escalate despite the combination of multiple allergy medications taken daily. Our new allergist in Boston discovered Brooke was now allergic to more than ninety percent of the foods she'd been eating, those we thought were safe, based upon her last allergy testing and food challenges. In fact, one of her most significant allergens is one of the three oils in her all important life sustaining formula. Brooke's sleep study revealed she has severe obstructive sleep apnea and at night requires bipap, a machine that forces air into the lungs.

This is mito! This is what can and does typically happen, unfortunately! All I do know for certain is that we finally have specialists that can help us and understand. This is a big relief for our family! However, there are many more individuals and families out there who aren't so lucky, those who are still struggling and suffering needlessly on their own.

Through our families' continued efforts to help raise awareness of mitochondrial diseases, we agreed to an updated local newspaper article several months ago. I received a phone call from a local woman who felt mitochondrial disease might explain what she had been dealing with for almost twenty years. All of the local doctors had given up on her and labeled her as depressed or crazy. They refused to do anything for her. I gave her the names of mitochondrial doctors and explained what she needed to do to confirm whether or not what she had was mitochondrial disease. About a month ago, she called to let me know she received the results of her muscle biopsy. Mitochondrial disease was confirmed. This confirmed my hope of helping others. We are educating our new local doctor who now realizes she has at least one other patient that might have a mitochondrial disease. Initially things were not working out with our new local doctor, so after a few months, I decided it wasn't worth the added aggravation. We 'd go it on our own with only the help of our specialists in Boston. This was equally frustrating. Essentially forced to return to the new local doctor after Randi's hospitalizations, I unfairly unloaded all of my pent up anger and extreme frustrations of dealing with all doctors on this one doctor's shoulders. Thankfully she is a good caring person and willing to take on our family's medical care. Moving to Boston is our only other option and one of the first things we were asked to consider doing when we began with our new specialists in Boston. So as hard as it is, and as long as it may take for our new local doctor to learn, we must do our part to educate and be patient. This is how the tide is being turned. It's a very slow exacerbating process.

It's happening one person at a time. Someday, soon I hope, everyone will be aware of and educated on mitochondrial diseases!

www.ingramcontent.com/pod-product-compliance
Lightning Source LLC
Chambersburg PA
CBHW031829170526
45157CB00001B/239